THE PRINCE OF EVOLUTION

Peter Kropotkin's Adventures in Science and Politics

Lee Alan Dugatkin is Professor of Biology and a Distinguished University Scholar in the Department of Biology at The University of Louisville. His books include *Mr. Jefferson and the Giant Moose* and *The Altruism Equation*.

Printed by CreateSpace, an Amazon.com company
© 2011 by Lee Alan Dugatkin

Dugatkin, L.A., 1962 –
The Prince of Evolution: Peter Kropotkin's Adventures in Science and Politics/Lee Alan Dugatkin

ISBN-13 978-1461180173
ISBN-10 1461180171
LCCN: 2011908284

THE PRINCE OF EVOLUTION

Peter Kropotkin's Adventures
in Science and Politics

Lee Alan Dugatkin

For Jerram Brown who believed in me when there was little reason to.

≋ TABLE OF CONTENTS ≋

Preface: The Inner Springs of Human Society ix

Chapter 1: Ex-Prince Peter 1
Chapter 2: Off to Siberia 10
Chapter 3: No Serious Organic Disturbances
in Peter 21
Chapter 4: The Ants have not Read Kant 33
Chapter 5: Strange Bedfellows 50
Chapter 6: We are all Brethren 62
Chapter 7: A Well-Preserved Russian Male 72
Chapter 8: The Old Fogey 83

Acknowledgments 96
Notes 97
Index 120

≡ PREFACE ≡

THE INNER SPRINGS OF HUMAN SOCIETY

"...{He is} that beautiful white Christ which seems to be coming out of Russia... {one} of the most perfect lives I have come across in my own experience."

Oscar Wilde[1]

Oscar Wilde was not the sort of man prone to effusive compliments. Who could possibly have merited such glowing praise from Wilde's typically satirical, razor-edged pen? That perfect life, the White Christ, belonged to a quite remarkable Russian scientist, explorer, historian, political scientist, and former prince by the name of Peter Kropotkin.

Kropotkin was one of the world's first international celebrities. In England he was known primarily as a brilliant scientist, but Kropotkin's fame in continental Europe centered more on his role as a founder and vocal proponent of anarchism. In the United States, he pursued both passions. Tens of thousands of people followed ex-Prince Peter—and that is how he was often billed— during two "speaking tours" in America.

Kropotkin's path to fame was unexpected and labyrinthine, with asides in prison, breathtaking 50,000-mile journeys through the wastelands of Siberia, and banishment, for one reason or another, from most respectable Western countries of the day. In his homeland of Russia, Peter went from being Czar Alexander II's favored teenage page, to a young man enamored with the theory of evolution, to a convicted felon, jail-breaker and general agitator, eventually being chased halfway around the world by the Russian Secret police for his radical—some might (and did) say enlightened—political views.

Both while in jail, and while on the run when he was entertaining and enlightening huge crowds, Kropotkin found the energy and concentration to write books on a dazzling array of topics: evolution and behavior, ethics, the geography of Asia, anarchism, socialism and communism, penal systems, the coming industrial revolution in the East, the French Revolution, and the state of Russian literature. Though seemingly disparate topics, a common thread—the *scientific* law of mutual aid, which guided the evolution of *all* life on earth—tied these works together. This law boils down to Kropotkin's deep-seated conviction that what we today would call altruism and cooperation—but what the Prince called mutual aid—was the driving evolutionary force behind all social life, be it in microbes, animals or humans. Traveling around the world, and trying to elude the Secret Police, simply gave Kropotkin the time, material and experience to develop his ideas.

Peter's theory of mutual aid came to him in the most unlikely of places. To follow in the footsteps of his hero, Alexander von Humboldt, when he was twenty years old, Kropotkin began a series of expeditions in Siberia. At that point, he was already an avowed evolutionary biologist—one of the few in Russia—and a great admirer of Darwin and his theory of natural selection.

50,000 miles later, and five years the wiser, Kropotkin left Siberia a Darwinian. But he was a very different kind of evolutionary biologist: a new species of sort. For in Siberia, Kropotkin had not found what he had expected to find.

Though still in its early gestation period when Kropotkin began his journey through Siberia, evolutionary theory of the day advanced that the natural world was a brutal place: competition was *the* driving force. And so, in the icy wilderness, Peter expected to witness nature red in tooth and claw. He searched for it. He studied flocks of migrating birds and mammals, fish schools, and insect societies. What he found was that competition was virtually nonexistent. Instead, in every nook and cranny of the animal world, he encountered mutual aid. Individuals huddled for warmth, fed one another, and guarded their groups from danger, all seeming to be cogs in a larger cooperative society. "In all the scenes of animal lives which passed before my eyes," Kropotkin wrote, "I saw mutual aid and mutual support carried on to an extent which made me suspect in it a feature of the greatest importance for the maintenance of life, the preservation of each species and its further evolution."[2]

Kropotkin didn't limit his studies to animals alone. He cherished his time in peasant villages, with their sense of community and cooperation: in these small Siberian villages, Kropotkin began to understand "the inner springs of the life of human society."[3] There, by observing "the constructive work of the unknown masses,"[4] the young scientist witnessed human cooperation and altruism in its purest form.

The conflict then arose in trying to align his observations with Darwinian theory. While he might easily have abandoned evolutionary thinking altogether, joining many other Russian scientists in dismissing Darwin's ideas as nothing more than Victorian smoke and mirrors, Kropotkin understood that evolutionary thinking could explain the diversity of life he saw around him. And so he set up the tightrope on which he would balance for the rest of his life. He advocated that natural selection was the driving force that shaped life, but that Darwin's ideas had been perverted and misrepresented by British scientists. Natural selection, Kropotkin argued, led to mutual aid, not competition, among individuals. Natural selection favored societies in which mutual aid thrived, and individuals in these societies had an innate predisposition to mutual aid because natural selection had favored such actions. Kropotkin even coined a new scientific term—*progressive evolution*—to describe how mutual aid became the *sine qua non* of all societal life—animal and human. Years later, with the help of others, Kropotkin would formalize the idea that mutual aid was a biological law, with many implications, but the seeds were first sown in Siberia.

From the Siberian tundra, Kropotkin's thinking turned to the political implications of mutual aid. The ants and termites, the birds, the fish and the mammals were cooperating in the absence of any formal organizational structure—that is without any form of "government." The same was true in the peasant villages, where mutual aid abounded, but a centralized government structure was nowhere to be seen. Kropotkin sensed great similarities with the writings of anarchists, which he had taken to covertly as a teenager. Leave people with complete freedom and autonomy, Peter had read in the anarchist literature, and they will naturally cooperate. In Siberia, Kropotkin had discovered this to be true not only for humans, but for all species that lived in groups. What marked so much in the natural world could surely help in politics and society.

"I lost in Siberia," Kropotkin would write," whatever faith in State discipline I had cherished before: I was prepared to become an anarchist."[5] Peter became so convinced that his scientific findings on mutual aid explained the biological underpinnings of political anarchy, that years after his trek through Siberia, he wrote in his obituary for Charles Darwin that, properly understood, Darwin's theories were "an excellent argument that animal societies are best organized in the communist-anarchist manner."[6] In time, Kropotkin's ideas on the science of mutual aid would lead to his rise as the most famous anarchist of his day.

Kropotkin today retains his moniker as a key founder of anarchist principles. And for more than 80 years—until about the 1960s—Kropotkin's ideas on mutual aid played a prominent, critical role in the study of behavior and evolution. And during that same period, the Prince's book-length treatments on ethics, geology, history and literature had a huge impact not only on those fields, but on areas as diverse as city planning, communist ideology and the modern "green" movement.

But we are getting ahead of our story. We need, first, to transport ourselves back to the Old Equerries section of Moscow, circa 1842, and learn more about the forces that shaped Prince Peter Kropotkin, descendant of the once powerful Rurik Dynasty.

≋ CHAPTER 1 ≋

EX-PRINCE PETER

Kropotkin, comrade, you have cleansed long since
Your genial name from "Prince"

From a poem written upon Peter Kropotkin's death[1]

Catherine the Great once described Moscow as "a seat of sloth...the nobles who live there are excessively fond of the place and no wonder: they live in idleness and luxury."[2] In the middle of the 19th century, this seat of sloth was divided into four quarters, each with its own distinctive personality. The winding streets of The Old Equerries Quarter were home to the nobles and demi-nobles, their mansions the scenes of late night revelries.[3] And in one of the green-shingled, six-columned homes in The Old Equerries Quarter, on December 21, 1842, during the reign of Czar Nicholas I, Peter Alexeivich Kropotkin was born.[4]

The Kropotkins were princes, Peter included. His ancestors were members of what was known as the Rurik Dynasty and had been rulers in the area of Smolensk.[5] The family could trace its lineage back to the early 1500s, when Prince Dmitri Vasilevich was given the nickname "Kropotka."[6] Peter's father, Alexander Petrovich Kropotkin, made sure that all who entered their home knew this history. "Our father," Peter wrote, "was very proud of the origin of his family, and would point with solemnity to a piece of parchment which hung on a wall of his study. It was decorated with our arms... ."[7] Alexander was a military man, an officer of the Czar's army, but he had seen little in the way of battle[8]—Peter once claimed he doubted his father had spent "a single night of his life at a bivouac fire,"[9] believing him to be more enamored with his uniform and "presenting arms" on command, than with an actual cavalry charge.[10] Stern, and lacking any sense of humor, Alexander ran his home like a regimental headquarters, with no interest in the intellectual or artistic—he was ashamed that his own brother was a published poet[11]—aspiring to military careers for each of his three sons.

Peter's mother, too, was from a military family, but Ekaterina Sulima had other passions as well. Her diaries, found by Peter long after her death, were filled with Russian verses, prohibited by censorship, as well as banned French and British literature that she had copied by hand. Though he was merely three years old when his mother died of consumption, Peter wrote of his childhood as "irradiated by her memory."[12] Peter's older siblings, Nicholas, Yelina and Alexander (often referred to as Sasha) also spoke glowingly of Ekaterina and her influence on their early lives.

A few years after Ekaterina's death, Peter's father married Elisabeth Karadino, who Peter referred to as his "cursed stepmother." All traces of Peter's mother were banished from the house, and the family shortly thereafter moved. But with them traveled his father's lavish lifestyle. Twelve hundred serfs were linked to the Kropotkin properties, fifty servants in the main house alone. "Four coachmen to attend a dozen horses," Peter recalled, "three cooks for the masters and two more for the servants, a dozen men to wait upon us at dinner-time ... and girls innumerable in the maid-servants' room...Dancing-parties were not infrequent...obligatory balls every winter...."[13]

Alexander and Elisabeth relished little above hosting grand events, which they also attended with fervor. When Peter was eight years old, there was a grand ball held in Moscow to commemorate Czar Nicholas' 25th year in power. Alexander and Ekaterina where invited to the ball, and Peter too attended, in place of a young boy too ill to play a Persian prince. Peter fit the costume perfectly, and joined sixty other children as flag bearers at the ball. The Czar was enchanted by young Kropotkin, who was carrying the flag of the Astrakhan region. At the imperial platform, Peter stood "amidst generals and ladies looking down upon me with curiosity."[14] The Czar then announced to one of his colleagues, "That is the sort of boy you must bring me."[15] Seven years later, Peter would join the Czar's Corps of Pages—one of the highest honors a young man could receive in Imperial Russia. At the time, though, a perplexed and frightened eight-year-old Peter cried out, "I am not a page. I will go home,"[16] just wanting to leave the ball and bring his brother Sasha some biscuits he had taken for him.

During the seven years between the Czar's 25th anniversary ball and Peter's entrance into the Corps of Pages, the Kropotkin children were taught by a French tutor, Poulain, and a Russian student named Smirnoff. Poulain, who taught them geography, mythology, history, and French, established order with a birch rod, and would parade the children to Alexander and Elisabeth's breakfast table each morning to recite their morning greetings in French. But, after Peter's sister Yelina complained to their father of their tutor's readiness to apply the birch rod for the slightest infraction, Poulain changed his ways and became, for Peter, "a lively comrade instead of a gruesome teacher."[17]

During these years, each summer the Kropotkins would pack their elaborate horse-drawn carriages and travel for days to their summer estate in Nikolskoye. Peter was an insatiable reader, and his bags at the time were packed with Alexander von Humboldt's *Cosmos*, and similar readings. On the long, but beautiful, trek to the summer house, Peter became a budding natural historian. His sorties into the woods, with its arctic foxes and reindeer roaming about under the dense canopy of summer, and his walks along the brooks filled him with joy, and imbued within him a passion for studying the natural world. "In that forest," he would write, "my first love of nature and my first dim perception of its incessant life were born."[18]

The emergent naturalist was also being schooled in the world of radical politics. His Russian tutor, Smirnoff, was a student at Moscow University, and introduced the Kropotkin children to contraband political material on nascent Russian movements in anarchism and socialism, even going so far as taking Peter on walking tours of the homes of writers and political agitators in Moscow.[19] After his French tutor, Poulain, shared stories of noblemen renouncing their titles during the French Revolution, Peter ceased to refer to himself as Prince, henceforth signing his name as *P. Kropotkin*.

Following his introduction to socialism and anarchism, Peter spent some time at the Moscow Gymnasium,[20] and in 1857, at age 15, he was admitted to the Corps of Pages, which was an academy that prepared the best and the brightest for service to the Czar, usually as military officers. The path it promised to a prestigious government position in Russia held little appeal for Peter, but he had no choice in the matter.[21]

Peter found the life of a page monotonous. He would write of horrible boredom in his diaries and letters to family. "I hate everything more with each day," Kropotkin wrote his brother, "… there is no one with the same inclinations I have."[22] He made few friends in the Corps, and the classes he took offered little of interest or challenge. "Day after day passes," Peter wrote Sasha, "almost the best days of life and you can't make use of them, you simply vegetate, you don't live."[23]

Plagued by his boredom and inspired by his developing sense of antipathy toward authority, Peter began acting out. In one case, he told a mathematics professor that he disliked that he would never open a book in his course and still obtain the highest grade. In a second instance, in response to an arbitrary exercise of power by one of his instructors, Kropotkin organized the other students in his class to bang their rulers on the table in protest, in an attempt to force the teacher out of the class. This was not the sort of action the Corps of Pages could tolerate from Russia's future military leaders, and Kropotkin was placed in solitary confinement in a dark cell for ten days. He long wore that confinement as a badge of honor.

Kropotkin's five years in the Corps did not pass without some pleasure. Peter ignored the Corps rule that "no students are permitted to have books or to subscribe to journals without the previous permission of the inspector of classes,"[24] and immersed himself in books on history, politics, philosophy, science and poetry. The other pages began calling him "father literature" and "the historian."[25] On Saturday evenings and Sundays Peter would bury himself in the library at his sister's home, and read through Voltaire's *Dictionnaire Philosophique* and the works of the Stoics. He composed his own course on medieval history, translated Voltaire into Russian, and wrote a physics textbook that was actually used in the Corps of Pages.

In the letters exchanged between Peter and Sasha during Peter's time in the Corps, Kropotkin the thinker and Kropotkin the scientist began to emerge.[26] Peter idolized his older brother, with whom he developed a sort of intellectual sibling rivalry. Peter accused his brother of falling under the spell of Kant, which he would have no part of: "…neither then nor later on, when we used to spend hours and hours in discussing Kant's philosophy, could my brother convert me to become a disciple."[27]

Sasha introduced his younger brother to Darwin's ideas soon after *The Origin of Species* was published in 1859. He also wrote to Peter of a professor of zoology at Moscow University, who had given three lectures on "transformism." Sasha was fascinated with the variability of species, and shared his fascination with Peter. Both would read *On The Origin of Species,* and their discussion about Darwin and evolution continued for many years thereafter. Peter's interest in evolution, his love of nature, and his passion for literature fused. Later he would write of a never-ceasing universe "... which I conceived as life and evolution, became for me an inexhaustible source of higher poetical thought, and gradually the sense of Man's oneness with Nature, both animate and inanimate—the Poetry of Nature— became the philosophy of my life."[28]

Peter's passion for science was not confined to evolution. Pages were often assigned tasks associated with surveying—particularly with respect to potential sites for forts. In making plans for lakes, roads and parks, Peter was introduced to the tools of a geologist. For him, the solitary time spent surveying provided a deep sense of enjoyment. "The independent work, the isolation under the centuries-old trees, the life of the forest which I could enjoy undisturbed," he wrote, "all these left deep traces on my mind."[29]

The 1850s were a good time for a young Russian man to turn to science. Though a political and military disaster,[30] the Crimean War catapulted science to the forefront of military and government agendas. And, coincidentally, Russians in the mid 19th century were also awaking to the inherent power of the scientific method. "Numbers of excellent books were published at that time in Russian translations," Peter wrote, "and I soon understood that whatever one's subsequent studies might be, a thorough knowledge of the natural sciences and familiarity with their methods must lie at the foundation."[31] Kropotkin and a few of his fellow pages were excited enough about all this that they built a makeshift laboratory in their dorm room.[32]

In June 1861, despite his occasional act of anti-authoritarianism, Peter's academic achievements earned him the position of Sergeant of the Pages. The highest honor that could be bestowed on a page, the Sergeant was the page de chambre of the Czar. "To be personally known to the Emperor," Peter understood, was "a stepping stone to further distinctions," including a "court life {that} has undoubtedly much that is picturesque about it." Peter felt little for the ordinary trappings of this honor; rather to him, the Sergeant position freed him from "the drudgery of the inner service of the school…" and came with private living quarters, which would provide a quiet space to read his books on history, science, geography and politics.

When Peter first became Sergeant of the Corps, he admired the Czar. Only months earlier, in February 1861, Alexander II decreed the imminent emancipation of the serfs. Kropotkin found Alexander II to be a different species of leader, genuinely concerned with the people—serfs, peasants or factory workers. Peter wrote of him "as a sort of hero; a man who attached no importance to the court ceremonies, but who, at this period of his reign, began his working day at six in the morning, and was engaged in a hard struggle with a powerful reactionary party in order to carry through a series of reforms, in which the abolition of serfdom was only the first step."[33] He even mused, "Imagination often carries a boy beyond the realities of the moment, and my frame of mind at that time was such that if an attempt had been made in my presence upon the Czar, I should have covered him with my body."[34]

But his reverence was fleeting. From his Sergeant perch, Kropotkin observed in the daily machinations of palace life the same despotic tendencies that characterized many rulers. The Czar was prone to a tempestuous nature, and held many in his court in sheer contempt. And the Czar's beneficence in freeing the serfs proved to be duplicitous, at least in Peter's eyes. Under the new system, ex-serfs were paid next to nothing, and now had no land to cultivate. The abolition actually proved financially lucrative for former feudal lords. For Peter, a critical component of any successful emancipation was education. When the State closed Sunday schools—one of the few places peasants had to practice reading—Kropotkin joined a group that tried to provide education to the poor. School closings were just the sort of arbitrary, mindless exercise of authority that Kropotkin was coming to despise, and he was both "ashamed and angry at the chicanery with which the peasants were defrauded of their moral rights after the liberation."[35] The aura that initially surrounded Alexander II had faded.

As Peter's term as a page came to an end in the Spring of 1862, most of his contemporaries longed for positions in "Her Majesty's Cuirassiers," but Kropotkin was not one for a life of parades and balls. He wanted to study at the university, but knew his father wouldn't oblige. So he sought an appointment that would have a nominal military component, and provide him with a chance to pursue his passions in science, history and politics.

He set his sights on Siberia, the newly annexed region of Amur.[36] In its terrain Peter found a "Mississippi of the East, the mountains it pierces, the subtropical vegetation of its tributary... the tropical regions which Humboldt had described."[37] The Amur became Peter's Galapagos Islands, the expedition that formally launched his natural history pursuits. Because it was just annexed, politics in the area was still in its infancy, and this region could also sate Peter's appetite for politics and history.

To the amazement of his peers, and to the consternation of his father and his brother, Peter requested an appointment to a mounted Cossack infantry in the Amur region. One of his favorite teachers in the Corps begged him to reconsider, pleading with Peter to turn to the university instead. His infuriated father telegraphed the director of the Corps, informing him that his son was forbidden from going to Siberia. The director then passed this information on to the Grand Duke (The Czar's brother), who was in charge of the Corps of Pages.

Amur seemed an unlikely destination until The Great Fire of St. Petersburg, when Kropotkin was credited with saving an entire barracks of the Corps of Pages from almost certain death. The Grand Duke summoned Kropotkin for his bravery and asked why he wished to go to Siberia. Persuaded by Peter's response, the Grand Duke freed him to go, forcing his father to acquiesce. Only one hurdle remained, that of the Czar himself. Alexander II asked his page de chambre, "So you go to Siberia?... Are you not afraid to go so far?" The young Kropotkin's reply must have stunned the Czar. "No, I want to work," Peter said. "Well, go," Alexander II told him, "one can be useful everywhere."[38]

On July 27, 1862, just a little over a month after graduating from the Corps of Pages, Kropotkin began a 50,000-mile trek through Siberia—his colder version of *The H.M.S. Beagle* voyages.

≡ CHAPTER 2 ≡

OFF TO SIBERIA

"Science is an excellent thing. I knew its joys and valued them...."

Peter Kropotkin[1]

Twenty-year-old Peter Kropotkin left St. Petersburg with few regrets. "I have finally gotten myself out," he wrote in his journal, noting that the only thing that he would miss of St. Petersburg was the Bolshoi Theatre.[2] Though Peter had demonstrated incredible resilience in his ascent in the Corp of Pages, he had little idea of the trials that awaited him. Reflecting on this time, he later wrote, "the five years that I spent in Siberia were for me a genuine education in life and human character."[3]

His passions lead him to Siberia, but he brought with him too his duties as a member of the Czar's army. His first mission was a report on the prisons of the Amur region, which aligned well with his budding political agenda. And to prepare for this unknown terrain, as always, Peter turned to books. Reading material was not easily obtained in Irtursk, the capital of East Siberia, but he managed to acquire papers published by the ministries of justice. Armed with this information, he went to the field with fervor. Not only did Kropotkin visit the local prisons, but he also examined "lockups," where prisoners were kept for a night or two on their path to exile in Siberia.[4]

Kropotkin found conditions in the prisons appalling, on par with what he read in Dostoevsky's *Buried Alive*. The structures were rotting, the people running them were contemptible bureaucrats, and the prisoners were treated as subhumans. Kropotkin would one day write that the border to Siberia should bear the inscription from Dante's *Inferno*, "*Abandon Hope All Ye Who Enter Here.*"[5]

His own hope was not so easily abandoned. Kropotkin worked tirelessly to produce a report proposing a more humane prison system. Though he shared it with Siberian officials, their lack of response infuriated young Peter. Kropotkin's experience with the State up to now had been through his personal interactions at the Palace of Alexander II. He already had his doubts about the centralization of power, and the lack of interest in improving prisons fueled his skepticism of all things State. He now saw government as prone to "insane paroxysms."[6] And though the administration in Siberia had good intentions, it was still a centralized administration that was inefficient and corrupt, largely, Peter believed, because a bureaucrat will always be concerned, first and foremost, with pleasing his superiors, not serving the people.

Government, Peter grew convinced, was the source of, not the solution to, the problems in Siberia, and perhaps Russia in general. This conviction grew stronger upon meeting exiled radical poet M.L. Mikhailov—an early advocate for women's rights.[7] Mikhailov, whose trial caused quite a stir in Czarist Russia,[8] gave Peter a copy of Pierre-Joseph Proudhon's anarchist tract *Système des Contradictions Economiques*. Although Kropotkin was already familiar with anarchist literature, and he would meet other political radicals during his time in Siberia,[9] Mikhailov's gift of Proudhon's tract came at a critical juncture in young Peter's life. He had experienced first-hand, through his work on prison reform, government ineptness and its effect on the lives of commoners.

Of his five Siberian expeditions, the most fruitful was a long journey ostensibly to find a path between the gold mines of Yakútsk and Transbaikália.[10] Kropotkin found the route, but the journey also revealed the extent and power of mutual aid—cooperation—in nature.

In winter, Siberia can be one of the most brutal places on the planet. Though Kropotkin tended to play down the severity of its climate, even he found it difficult "lying full length in the sled…wrapped in fur blankets, fur inside and fur outside…when the temperature is forty or sixty degrees below zero, Fahrenheit."[11] Temperature aside, the Siberian environment can be uncompromising. Once, after a mission involving shipping supplies to a famine-stricken community along the Amur River, Peter had the chance to briefly visit his brother Sasha in St. Petersburg. The 3200-mile trek from Siberia entailed many broken wheeled carts and the crossing of icy rivers. Sometimes, Kropotkin would recruit local peasants to help make such crossings. At the Tom River, the peasants refused, unless Peter would indemnify them and sign a receipt attesting that "I, the undersigned, hereby testify that I was drowned by the will of God, and through no fault of the peasants."[12] Kropotkin agreed, and a young peasant boy led the way.

Few trained explorers would have survived Peter's ordeals in Siberia. His sojourns to various parts of this region included a "two thousand mile journey in a row-boat, changing rowers at each village, every twenty miles or so."[13] But Kropotkin possessed incredible fortitude. His trials and tribulations in Siberia—"fifty thousand miles in carts, on board steamers, in boats, but chiefly on horseback"[14]— by his own reckoning, only added to this vigor. He learned "how little man really needs as soon as he comes out of the enchanted circle of conventional civilization."[15]

In this land of ice and snow, Kropotkin saw mutual aid at work everywhere. When Peter set off for Siberia, he expected to encounter nature red in tooth and claw, the world he and Sasha spoke of in their discussions about evolution and natural selection. "I failed to find, although I was eagerly looking for it," he wrote, "that bitter struggle for the means of existence, among animals belonging to the same species, which was considered by most Darwinists (though not always by Darwin himself) as the dominant characteristic of the struggle for life, and the main factor of evolution."[16] Instead, Peter saw the real struggle to be not between individuals of the same species, but "the struggle for existence which most species of animals have to carry on *against an inclement Nature*."[17] The severity of Siberia itself was driving the evolutionary process.

From the millennia-long struggle against inclement conditions, mutual aid had evolved. Animals cooperated to combat the harsh realities of Siberia. Mutual aid, in Kropotkin's eyes, was foundational to animal life: "Wherever I saw animal life in abundance," Peter noted "... on the lakes where scores of species and millions of individuals came together to rear their progeny; in the colonies of rodents; in the migrations of birds which took place at that time on a truly American scale along the Usuri; and especially in a migration of fallow-deer which I witnessed on the Amúr, and during which scores of thousands of these intelligent animals came together from an immense territory...in all these scenes of animal life which passed before my eyes, I saw Mutual Aid and Mutual Support carried on."[18]

Kropotkin chronicled a vast array of cooperative behaviors and became convinced that the depiction of the animal world as "a pitiless inner war for life"[19] was simply wrong. Despite what

Darwinists were saying, to acquiesce to such a view was "to admit something which not only had not yet been proved, but also lacked confirmation from direct observation,"[20] which Kropotkin was incapable of.

Peter's scientific training drew him away from emotionally laden terms such as "love" to describe the cooperative behavior he observed: "It is not love, and not even sympathy..." he wrote, "which induces a herd of ruminants or of horses to form a ring in order to resist an attack of wolves; not love which induces wolves to form a pack for hunting; not love which induces kittens or lambs to play, or a dozen of species of young birds to spend their days together in the autumn." Such actions he attributed to the practice of mutual aid, which he found to be "infinitely wider than love or personal sympathy—an instinct that has been slowly developed among animals and men in the course of an extremely long evolution."[21]

Though life in all human towns and settlements in Siberia was difficult, mutual aid thrived in such localities. Peter found that mutual aid there correlated with government interference. In those towns and settlements near cities, with their cadres of bureaucrats, mutual aid was suppressed by government meddling. "The years that I spent in Siberia," Kropotkin wrote, "taught me... the absolute impossibility of doing anything really useful for the mass of the people by means of the administrative machinery. With this illusion I parted forever."[22]

But human mutual aid did flourish in parts of Siberia. In those towns and villages farther from population centers, mutual aid thrived. Here, freed from the actions of bureaucratic bunglers, Siberian peasants displayed almost unbounded mutual aid. "The constructive work of the unknown masses," wrote Peter, "which so seldom finds any mention in books...the importance of that constructive work in the growth of forms of society, fully appeared before my eyes...to see the immense advantages which they {communities} got from their semi-communistic brotherly organization, and to realize what a wonderful success their colonization was, amidst all the failures of state colonization, was learning something which cannot be learned from books."[23] Kropotkin would eventually write prolifically of human mutual aid,[24] but for the time being, his observations produced "floods of light which illuminated...subsequent reading."[25]

His brother Sasha held to a very different view of human mutual aid than Peter. In his letters to Peter, Sasha argued that what appeared to be altruism and mutual aid was just another manifestation of self-interest. All mutual aid was just a byproduct of greed and ego.[26] The brothers Kropotkin had drifted further apart philosophically: "It is not love of my neighbour," wrote Peter, "whom I often do not know at all, which induces me to seize a pail of water and rush towards his house when I see it on fire; it is a far wider, even though more vague feeling or instinct of human solidarity and sociability which moves me. So it is also with animals."[27] Indeed, his observations of both humans and animals had drawn him to a startling and dramatic conclusion: mutual aid was not only common, but "of the greatest importance for the maintenance of life, the preservation of each species, and its further evolution."[28] He would soon speak of mutual aid as a biological law.

In Siberia, Peter's scientific interests were not limited to natural history and evolution. The training that he had received as a page prepared him well for the geographical expeditions he undertook in Siberia. Often uninspired by administrative tasks, Kropotkin perpetually sought new opportunities. And, when he was offered the chance to "undertake geographical explorations in Manchuria,"[29] he quickly accepted.

The mission's goal was to map out a way through the mountains to the great Amur River. The region was shrouded in folklore, and the expedition had its risks, including an ongoing border dispute between Russia and China over the very area they were to explore. "The temptation to visit a country which no European had ever seen {was} too great for an explorer to resist,"[30] Peter wrote, and so to avoid diplomatic problems he posed as a trader, traveling under the pseudonym Petr Alexeiev. He and his team were successful, discovering a path through the mountains: the discovery promoted Kropotkin's first published science paper in *Memoirs of the Siberian Geographical Society*.

Word of Peter's skills spread quickly, and soon he led other geographical expeditions, most of which were through mountainous terrain. Kropotkin was fascinated by the geology of mountains, and from his own expeditions, and by integrating the findings of other geographers, he became obsessed with determining the orientation of the mountain ranges in Asia.[31] For centuries people, including his hero, Alexander von Humboldt, had argued that the orientation of the giant mountain ranges in Asia was either north-south or east-west. "One day, all of a sudden," Kropotkin wrote, "the whole became clear and comprehensible, as if it were illuminated with a flash of light."[32] This flash of light led Kropotkin to his most famous geological finding: "The main structural lines of Asia," Peter declared, "are not north and south, or west and east; they are from the southwest to the northeast, just as, in the Rocky Mountains and the plateaus of America, the lines are northwest to southeast... ."[33] But Kropotkin was not satisfied with description alone, he was also interested in causation: why were the mountain ranges of Asia oriented as they were? And he found his answer: "the mountains of Asia are not bundles of independent ridges, like the Alps... high border ridges have towered up along its fringes, and in the course of ages, terraces, formed by later sediments, have emerged from the sea, thus adding on both sides to the width of that primitive backbone of Asia."[34]

Peter considered this discovery his "chief contribution to science,"[35] and described it in poetic terms: "What has seemed for years so chaotic, so contradictory, and so problematic takes at once its proper position within an harmonious whole...He who has once in his life experienced this joy of scientific creation will never forget it; he will be longing to renew it; and he cannot but feel with pain that this sort of happiness is the lot of so few of us."[36]

Five years in Siberia shaped Kropotkin not only as an evolutionary biologist and a geographer, but also as an emergent political activist. "I lost in Siberia," he would write, "whatever faith in state discipline I had cherished before. I was prepared to become an anarchist."[37] This loss of faith was, in part, a result of Mikhailov's gift of Proudhon's anarchist tract, Kropotkin's interactions with other political "radicals," and his daily frustrations with the ability of government to serve people. He had witnessed in the towns and villages of Siberia that government actually worsened the lives of people. "I was brought into contact with men of all descriptions," Kropotkin wrote, "the best and the worst; those who stood at the top of society and those who vegetated at the very bottom... I had ample opportunities to watch the ways and habits of the peasants in their daily life and still more opportunities to appreciate how little the state administration could give to them, even if it was animated by the very best intentions."[38] But there was more to Kropotkin's anarchist beliefs than this; they evolved naturally from his work on mutual aid in animals.

The political philosophy of anarchism posits that no centralized government is necessary for people to lead happy, just and equitable lives. Leave people to themselves, anarchists argue, and they will treat one another with respect and decency. As Kropotkin himself did, anarchists "arrive at the conclusion that the ultimate aim of society is the reduction of the functions of government to nil—that is, to a society without government, anarchy."[39]

The mutual aid that Kropotkin found among people in all corners of Siberia legitimatized this view. But it was the mutual aid that Peter observed in animals that made him a true anarchist. The fact that animals displayed mutual aid—and did so in the absence of anything remotely like a government—suggested its deep biological roots. Kropotkin felt that the process of evolution had favored mutual aid in animal populations, and if he had to put a political label on the way that these animals behaved, it would be "anarchy."

The link between anarchy and mutual aid in animals had both political and scientific consequences for young Peter. If animals cooperated in the absence of government, then it seemed incomprehensible to Kropotkin that humans could not find a way to break the shackles of government. Surely we could find a method for living free from "the fetters of the State."[40] Evolution, Kropotkin wrote, favored "aggregation{s} of organisms trying to find out the best ways of combining the wants of the individuals with those of co-operation for the welfare of the species."

Political anarchists, Peter wrote, had biology on their side: they were only adopting "the course traced by the modern philosophy of evolution." Anarchy, he noted, was "a mere summing-up of…the next phase of evolution."[41] The link between science and politics became cemented into place as Kropotkin began to think that "anarchism represents more than a mode of action and a mere conception of a free society; that it is part of a philosophy, natural and social… that it must be treated by the same methods as natural sciences."[42]

Eventually Kropotkin left Siberia—the very place that had made him the man he had become. Why? Why leave a place so brimming with mutual aid, a place where one could learn of geology and politics and be left alone to do so? Ostensibly, it was a group of Polish exiles that prompted Peter's move. They were excavating a new road around Lake Baikál, and made a desperate attempt to escape, and to flee to China across Mongolia. The Czar's army quashed the attempt in typical military fashion and Peter felt it time to distance himself from all things Czarist. But even if the Polish mini-revolt had not occurred, Kropotkin was already prepared to depart. He was determined to get to the University and was now mature enough to do so without concern for his father's approval or financial support.[43]

In the Fall of 1867, Kropotkin took up mathematics at the University of St. Petersburg. But Kropotkin's intellectual appetite still could never be sated with a single subject: "A student of the mathematical faculty has, of course, very much to do, but my previous studies in higher mathematics permitted me to devote part of my time to geography; and, moreover, I had not lost in Siberia the habit of hard work."[44]

He used his time to learn more about the politics of anarchy. Kropotkin was intensely interested in the Franco-Prussian War, and the Paris Commune—a short-lived entity that emerged after France's loss in that war. This Commune, or city-council, reigned over Paris for two months during the Spring of 1871, and was made up of many different political factions, the most interesting to Kropotkin being the socialists and anarchists. Peter and Sasha went to meetings where they spoke of the coming revolution that would engulf Russia as a result of the Paris Commune.[45]

When the Paris Commune failed, many of its members emigrated to Switzerland. In February 1872, Peter requested a month waiver from the University to visit Zurich. Once there, he met with anarchists, socialists and communists. The trip grew to three months, with visits to Geneva and Jura, where the Jura Federation—the anarchist wing of the International Workingman's Association—was located. Peter soaked in the radical politics, and returned to St. Petersburg, via trains through Warsaw and Krakow, clandestinely bringing along with him newspapers that were prohibited by the censors.

Kropotkin's Swiss sojourn was only one measure of an atypical University of St. Petersburg student. He spent time translating Herbert Spencer's works on sociology, evolution and psychology into Russian.[46] And his work on the structural orientation of the mountains of Asia had already made him a minor celebrity in Russian geographical circles. Still a student, he became secretary to the physical geography section of the Russian Geographical Society, which asked him to join a planned expedition to the Arctic, and though he accepted, and was to pilot a Norwegian schooner (though he never had been to sea), financial woes trumped the expedition.[47] In its place, toward the end of his five years at the University of St. Petersburg, Peter was sent by the Russian Geographical Society on a tour of Sweden and Finland to explore their glacial deposits. He collected a mass of data which he envisioned filling the pages of a book on the region's physical geography.

This was to be more than a book on the geography of Finland and Sweden, it was to be a general and practical guide on how geography could improve the life of Russian peasants. But the more he contemplated this type of book, the less enamored he became with the idea. Such a book, Peter wrote, would have to say things like "Here an American stump-extractor would be invaluable; there certain methods of manuring would be indicated by science." That language would not do for Kropotkin: "what is the use of talking to this peasant about American machines, when he has barely enough bread to live upon from one crop to the next; when the rent which he has to pay for that boulder-clay grows heavier and heavier in proportion to his success in improving the soil?... How dare I talk to him of American machines, when all that he can raise must be sold to pay rent and taxes?"[48] Kropotkin, focused as he was on mutual aid, geography, and the politics of anarchy, would not write such a book. Instead, he would trade words for action.

≋ CHAPTER 3 ≋

NO SERIOUS ORGANIC DISTURBANCES IN PETER

> *"Here were annals of murder and torture, of men buried alive, condemned to a slow death, or driven to insanity in the loneliness of the dark and damp dungeons."*
>
> Peter Kropotkin[1]

During his time at the University, Kropotkin's ideas on evolution and cooperation—indeed, on science in general—were, to some extent, placed on the intellectual back burner. He needed time to digest all the wonders of mutual aid that he had been privy to in Siberia. And he was beginning to believe that most of the people that were doing science in Russia at the time were "philistine"[2] autocrats who were more concerned with preserving the government than advancing knowledge. Above all, Peter's passion for science was temporarily eclipsed by politics.

Kropotkin's liberal ideas were "in sorely bad repute[3]" with the Czar's government, and many of the leaders of the revolutionary anarchist movement were either in jail or in hiding. The State crackdown included stifling the educational system that Kropotkin believed needed to include a technical component which would allow people to provide for themselves. "All Russia wanted technical education, but the ministry opened only classical gymnasia," Prince Peter recalled, and "all boys promising to become something and to show some independence of thought being carefully sifted out... Technical education—in a country which was so much in want of engineers, educated agriculturists, and geologists—was treated as equivalent to revolutionism."[4]

This suppression further fueled Peter's hatred of the State. As an older student, he had few friends at the University, but found kindred souls in the "Tchaykóvsky Circle,"[5] which he came to view as family.[6] This circle was a growing group of underground radicals—mostly anarchists and nihilists—who were attempting to educate the peasants to the evils of the State[7]. At first, members of the Tchaykóvsky Circle, who insisted on "morally developed"[8] individuals as members, were leery of "Prince" Peter's aristocratic background. And, as with most groups of young radicals, they tended to inherently dislike anyone much older than they were. Once Kropotkin began to interact with them, however, Circle members quickly realized that Peter was not only "completely young in spirit,"[9] but that he was one of their own politically. They were keen on having someone with his scientific knowledge in the group.[10] In no time, Peter was not only a fervent group member, but was holding Circle meetings in one of the Kropotkin homes.[11]

Early on, one of the most alluring aspects of the Tchaykóvsky Circle for Peter was that the main means by which this group worked was by distributing books. The Circle operated a small press to print their own material, as well as already published books that their members had translated into Russian—books, described by Peter, as "the works of the philosophers, the writings of the economists, the historical researches of the young Russian historical school."[12] Always drawn to books and the power of education, Kropotkin welcomed the chance to stir up discontent with the government, and better the life of the working class.

As time passed, the Tchaykóvsky Circle began to move from book distribution to direct interactions with the peasants. "Young men went into the villages as doctors, doctors' helpers, teachers, village scribes, even as agricultural laborers, blacksmiths, woodcutters, and so on, and tried to live there in close contact with the peasants," Peter noted, and never the sexists, the Tchaykóvsky Circle "girls passed teachers' examinations, learned midwifery or nursing, and went by the hundreds into the villages…These people went without…any thought of revolution. They simply wanted to teach the mass of the peasants to read, to instruct them in other things, to give them medical help, and in any way to aid in raising them from their darkness and misery."[13]

The State did not sit idly with such agitators on the loose, and sought out members of the Tchaykóvsky Circle. Nikolai Tchaykóvsky himself—who was a chemist by training—was arrested on numerous occasions and spent months at a time in prison. Soon after he published an article in the *Bulletin of the Russian Academy of Sciences,* and was on the brink of completing his final University examinations, he was hauled off to jail. Though quickly released on this occasion, Tchaykóvsky was warned "If we arrest you once more, we shall send you to Siberia."[14] And if Tchaykóvsky didn't already understand the horrors that jail in Siberia would entail, Kropotkin would surely have filled him in.

As with all of his passions, Kropotkin immersed himself in the Tchaykóvsky Circle, meeting with peasants,[15] teaching them about the emerging labor movements in Western Europe, and imploring his audience to educate themselves. Peter dressed the part of the peasant often, and spread the words of the anarchists: that government was inherently an evil enterprise, and that individuals naturally cooperate and solve problems better than the State. All of this energized Kropotkin, producing: "that exuberance of life when one feels at every moment the full throbbing of all the fibres of the inner self, and when life is really worth living."

Peter was not only out amongst the people, but continued to spread the anarchist message in his favorite medium, the written word. He was deeply involved with a Tchaykóvsky Circle pamphlet—perhaps the groups' most influential writing—entitled *Emilian Ivanovich Pugachev, ili bunt 1773*. This pamphlet was a romanticized novel-like story of Pugachev's 1773 rebellion against Catherine the Great, and in it, Kropotkin used the genre of historical fiction to promote the anarchist ideal.

Anarchy was not the only subject of his writing. Peter felt bound to complete the report of his journey to Finland and other work for the Geographical Society. As he worked on his report, the police stepped up the arrest of political agitators. Under such circumstances, Kropotkin, like other members of the Tchaykóvsky Circle, would normally have gone underground, and hidden among the peasants. But Peter had a problem. He had promised to give a talk to the Russian Geographical Society on the extension of the Russian ice cap, and he felt obliged to honor that commitment.

The Geographical Society meeting came, and Peter's ideas on glaciers were so well received there that he was nominated to be President of the society. Kropotkin was honored, but was more concerned about whether he "should not spend that very night in the prison of the Third Section."[16] Immediately after the meeting, he returned home, and fearing arrest, destroyed everything that might have been compromising. The next day at dawn, a young friend of his—one of the servant girls who worked in the apartments where Peter lived—knocked on his door and warned him to flee. The Third Section was at the front doors. Peter fled in a coach, but the authorities quickly overtook him. Peter, who knew all too well the power of the Third Section, was going to prison. And everyone would be speaking of this remarkable turn of events: "Upper

Saint Petersburg," a contemporary account noted, "was extremely scandalized by the arrest."[17]

As the carriage carrying Kropotkin approached the Peter and Paul prison,[18] Kropotkin feared the worst. With a reputation on par with that of the Bastille, all in Russia had heard horror stories of Peter and Paul. "Here Peter I tortured his son Alexis and killed him with his own hand," wrote Kropotkin, "here the Princess Tarakánova was kept in a cell which filled with water during an inundation, the rats climbing upon her to save themselves from drowning...here were annals of murder and torture, of men buried alive, condemned to a slow death, or driven to insanity in the loneliness of the dark and damp dungeons."[111]

Every aspect of the Peter and Paul Prison was designed to break a prisoner's will. Immediately upon arrival, Kropotkin was "required to take off all my clothes, and to put on the prison dress—a green flannel dressing-gown, immense woolen stockings of an incredible thickness, and boat-shaped yellow slippers, so big that I could hardly keep them on my feet when I tried to walk." The outfit was meant to humiliate prisoners, but it had an especially harsh effect on Peter, who "always hated dressing-gowns and slippers." The thick stockings, he wrote, "inspired me with disgust."[19]

Kropotkin was locked away in a part of the prison that had been built to store casements and guns, not prisoners. Aware that his time in Peter and Paul was indeterminate in length, Peter closely inspected his new cell. All of his actions could be observed by prison authorities through a small slit that Kropotkin referred to as "the Judas."[20] The cell housed an iron bed, a small table and stool and a washstand. Though the room contained a single window, "the rays of the sun could never penetrate it; even in summer they were lost in the thickness of the wall."[21] Light was not the only luxury Kropotkin was denied—sound too was cut off: to muffle any noise from the outside, or from the few other prisoners—of seventy cells, only six were occupied[22]—the floor and walls of his cell were covered by felt. His food was pushed through a small grate in the door by a guard who had been instructed never to converse with the Prince. "Absolute silence," Kropotkin wrote, "reigned all round." In response, Peter began singing: often quite loudly.

Though incarcerated in the State's worse prison, Kropotkin still had much work to do, including completing his report on geography. And, perhaps, more importantly, he believed that some day he would be free, and when that day came, he would need to be physically and mentally fit. To keep his body sound, Kropotkin pledged to himself: "I will not fall ill…I will take plenty of exercise, practice gymnastics, and not let myself be broken down by my surroundings."[23] He developed a strict daily routine of exercise. "Ten steps from one corner to the other is already something," he wrote. "If I repeat them one hundred and fifty times, I shall have walked one verst {two thirds of a mile}. I determined to walk every day seven versts—about five miles: two versts in the morning, two before dinner, two after dinner, and one before going to sleep."[24] For strength and flexibility, he turned to the heavy stool in his cell: "I lifted it by one leg, holding it at arm's length. I turned it like a wheel, and soon learned to throw it from one hand to the other, over my head, behind my back, and across my legs."[25]

Keeping his mind sharp was a greater challenge. Most prisoners were either illiterate or had no desire to read while incarcerated. But Kropotkin always hungered for books, an appetite not dampened by his surroundings. The authorities allowed him to keep the first volume of George Lewes's series on physiology, and eventually more books would come. "I…even arranged for myself a treat on Christmas Eve," Peter wrote. "My relatives managed to send me then the Christmas stories of Dickens, and I spent the festival laughing and crying over those beautiful creations of the great novelist."[26]

Peter also implored jailors to give him paper and pen to write. They initially refused, as such an exception to prison policy would have to be granted by the Czar himself. Sasha began a campaign on his brother's behalf. Fortunately for Peter, the Geographical Society was still quite keen on Kropotkin finishing his work for them, and with their assistance, and the support of the Russian Academy of Sciences, Sasha was able to get the Czar's permission to let Peter write. The Czar's instructions were clear: Kropotkin could write each day, but only until sunset. Despite the arbitrary, and seemingly senseless, nature of the sunset cutoff, Peter "could hardly express...the immensity of relief I then felt at being enabled to resume writing." To have the chance to write, Kropotkin remarked, "I would have consented to live on nothing but bread and water, in the dampest of cellars."[27]

Peter's obsession turned to his report for the Geographic Society. This would not be the "how-to" geography book for peasants that he once envisioned before his trip to Finland, but a more classic science book. In it, he would make his second major contribution to the science of geography. To assist him in this endeavor, the Russian Academy of Sciences "put its admirable library at my service," Peter wrote, "... including the whole of the Swedish Geological Survey publications, a nearly complete collection of reports of all arctic travels, and whole sets of the Quarterly Journal of the London Geological Society."[28]

During his time in Siberia, Kropotkin had studied the scratched and polished boulders that he and his team of explorers found in the river valleys.[29] From the markings on the boulders and the ground, he hypothesized that these rocks were moved to the valleys by a giant glacier that must have covered the mountains long ago.[30] Kropotkin found evidence for this during his subsequent explorations in Finland. He gathered his ideas on glaciation together in a magnificent work, *Researches on the Glacial Period*, which contained more than 100 hand drawn maps.[31]

For two years Peter languished in the Peter and Paul Prison, with only the occasional visit from family. There was also an unexpected visit by the Grand Duke who came by to tell Kropotkin how disappointed he was that a former page would transform himself into an enemy of the State. During the latter part of Peter's prison time, Sasha was arrested. They eventually sent him to Siberia, but not before denying Peter's brother the chance to spend a few last moments with his young son who was dying of consumption. These events, combined with the fact that his case had still not officially gone before a court of justice, had a profound effect on Peter, and his health began to suffer. Emotionally and physically drained, Kropotkin became quite ill in "the close atmosphere of the tiny cell... the temperature changed from a glacial cold to an unbearable heat."[32] He was soon diagnosed with a severe case of scurvy: so severe that his sister Yelana pleaded with the warden to release him for health reasons. The warden told her "if you bring me a doctor's certificate that your brother will die in ten days, I will release him."[33] Yelana, though, was not able to secure such a note, as the warden insisted it come from a prison doctor. Instead the prison doctor "found no serious organic disturbances in Peter,"[34] but nevertheless suggested that the prisoner be moved to a prison hospital.

On May 24, 1876, Kropotkin was transferred to the Peter and Paul Prison hospital,[35] located on the outskirts of St. Petersburg. He knew of this "hospital," as two of his comrades had been sent there, but only after it was clear that they would die of consumption. But the situation was not nearly as grim as he expected. His cell was comfortable, and had a beautiful view of a quaint street with rows of trees. Local carpenters would often sing as they completed their work, providing some peace for Peter and his fellow prisoners. These amenities were ameliorated, though, by the placement of two sentries outside Kropotkin's door—an "honor", Kropotkin noted, that was reserved for prisoners of special note.

After a few days in the prison hospital, Kropotkin began to recover, breathing "the balmy air of May with a full chest."[36] He was allowed some contact with people outside, and he and his colleagues hatched an audacious plot to escape. The details took shape after one of the more sympathetic guards suggested that Kropotkin request to be let out for an hour walk each day in the yard. Peter's request was approved by the staff doctor, and then by the Minister of Justice. Peter took his daily constitutional from 4-5pm, a walk that was contaminated by "the heavy white smoke thrown off by the chimney of the Mint which overlooks the yard," which often "completely poisoned {the air} during easterly winds."[37]

As he walked, Kropotkin cased the surrounding area. The yard was enclosed, but each day the gate was opened to allow peasants to bring in carts full of wood for fuel. Peter knew immediately that this gate, and its daily opening, could be the key to freedom. He and his friends on the outside began planning an escape. So excited was the Prince that he had trouble writing his co-conspirators, since his trembling hand was "unable to use the cipher, as this nearness of liberty makes me tremble as if I were in a fever."[38]

Kropotkin and his colleagues took a month to plan the escape. This long period of preparation was discomforting, as Peter knew that if the authorities believed he was recuperating, he might be transferred back to the real prison, where escape was impossible at any time. Still, the plot, which involved "hundreds of unforeseen details which always spring up around such conspiracies,"[39] required much in the way of planning and organization, and Peter did his best to appear sickly, and extend his stay.

The original escape plot was to go down like this:

> "A lady is to come in an open carriage to the hospital. She is to alight, and the carriage to wait for her in the street, some fifty paces from the gate. When I am taken out, at four, I shall walk for a while with my hat in my hand, and somebody who passes by the gate will take it as the signal that all is right within the prison. Then you

must return a signal: 'The street is clear.' Without it I shall not start; once beyond the gate I must not be recaptured. Light or sound only can be used for your signal. The coachman may send a flash of light,—the sun's rays reflected from his lacquered hat upon the main hospital building; or, still better, the sound of a song continued as long as the street is clear; unless you can occupy the little gray bungalow which I see from the yard, and signal to me from its window. The sentry will run after me like a dog after a hare, describing a curve, while I run in a straight line, and I will keep five or ten paces in advance of him. In the street, I shall spring into the carriage and we shall gallop away. If the sentry shoots—well, that cannot be helped; it lies beyond our foresight; and then, against a certain death in prison, the thing is well worth the risk."[40]

Every other day for two weeks before the escape, Kropotkin's co-conspirators drove up to the prison hospital in their coach, and, eventually ceased drawing any attention.

The date of escape was set for June 29[41], 1876 and the plan remained unchanged, except that rather than light or sound being the countersignal, a red balloon would be launched to let Kropotkin know that his co-conspirators were ready.[42] When Kropotkin took his walk that day, and gave the signal, no red balloon was launched in response. Failure. Kropotkin soon learned why: "The impossible had happened that day. Hundreds of children's balloons are always on sale in St. Petersburg... That morning there were none...One was discovered at last, in the possession of a child, but it was old and would not fly. My friends rushed then to an optician's shop, bought an apparatus for making hydrogen, and filled the balloon with it; but it would not fly any better: {still} a lady attached the balloon to her umbrella, and, holding the umbrella high over her head, walked up and down in the street along the high wall of our yard; but I saw nothing of it... ."[43] What's more, the streets that the carriage would have traversed were blocked by peasant carts, so even had Kropotkin escaped from the yard, he would have soon been caught.

Not deterred, a new plan was conceived—one that added more spies to make sure the roads from the hospital prison were clear, and that detailed the use of a bungalow "headquarters" for organization. Conspirators recruited new assistants to distract police and passersby: one pretended to be in search of room and board, while another provided signals by sitting on the curb and spitting cherry pits either to the left or the right.[44] Kropotkin's colleagues snuck the plan to him on a tiny piece of paper concealed in a pocket watch that a young lady, Madam Lavrova, had brought to the prison for him that day. This new plan was even more daring than the one that failed—even the Prince, who did not scare easily, was "seized with terror, so daring was the feat."[45] This time the signal he awaited was a violin playing in the distance, instead of a red balloon.

At the assigned time of 4pm, Kropotkin began his walk. He heard the violinist[46] playing "a wildly exciting mazurka from Kontsky, as if to say, 'Straight on now — this is your time!'"[47] Bolting for the gate, Kropotkin could almost feel the breath of the guard who began to chase him. "The sentry was so near to me," Peter noted, that the guard "was so convinced that he could stop me in this way that he did not fire. But I kept my distance, and he had to give up at the gate."[48] Outside the gate was a carriage, with a driver that shouted "Jump in, quick, quick!"[49] But to do that, Peter had to get by one last guard posted outside the gate. Fortunately, his collaborators had planned for this problem, and one distracted the hapless, and perhaps drunk, guard by showing him the wonders of a microscope. This distraction worked to perfection, and Peter dove into the carriage and rode to freedom.

The Czar and the Third Section police were furious when they heard of the escape, and a huge manhunt was launched.[50] Though hundreds of copies of his picture were posted all over St. Petersburg, Kropotkin seemed unfazed. He first went to a barber shop to shave off his distinctive beard.[51] Next, as if to taunt those on the hunt for him, he dined at the famous Donon restaurant in St. Petersberg, following a friend's advice that "no one will ever think of looking for you at Donon... we shall have a dinner, and a drink too, in honor of the success of your escape."[52] The meal at Donon went off without a hitch. The Third Section searched frantically for Peter for weeks, but found nothing: soon after his escape, using the passport of a friend, Kropotkin crossed over to Finland and from there to Sweden, where he went to the port of Christiania and set sail on steamer ship to England.

≡ CHAPTER 4 ≡

THE ANTS HAVE NOT READ KANT

> *"The ant, the bird, the marmot… have read neither Kant nor the fathers of the Church nor even Moses…The idea of good and evil has thus nothing to do with religion or a mystic conscience. It is a natural need of animal races. And when founders of religions, philosophers, and moralists tell us of divine or metaphysical entities, they are only recasting what each ant, each sparrow practices in its little society."*
>
> Peter Kropotkin[1]

Kropotkin's ocean voyage to Britain was difficult, with poor weather and choppy seas. But that hardly mattered to Peter, who was reinvigorated with freedom and, more importantly, the chance to act: "After the two years that I had spent in a gloomy casemate," he wrote, "every fibre of my inner self seemed to be throbbing and eager to enjoy the full intensity of life."[2]

Upon arriving at Hull, England, Kropotkin immediately began to search for work. After a brief sortie to Edinburgh, Scotland, he returned to London, where, under the alias, "Mr. Levashoff," he landed a position as book reviewer for the up-and-coming science journal, *Nature*. Though the job would last, the alias would not. In a comical moment, J. Scott Keltie, an editor at *Nature*, asked Mr. Levashoff to review the geography books of one Peter Kropotkin.[3] Peter, honest to the core, confessed that he in fact was the said Kropotkin, and should recuse himself as reviewer. Keltie, who had read of Peter's prison escape, told Kropotkin to review his own books using his alias, and "simply tell the reader what the books were about."[4]

While working at *Nature,* Kropotkin continued his geographical writings. What was missing for the ex-Prince, now expatriate, was the chance for direct action in politics, as the anarchist movement had not yet fully emerged in England. As soon as he found geographical work he could do in Switzerland, Peter moved there, where he reacquainted himself with the Jura Federation, and settled in the village of La Chaux-de-Fonds. He struck up a friendship with some of the leading anarchists of the day, including Elisee Reclus, who like Peter, was a geographer. It was as part of the Jura Federation that Peter penned one of his most famous essays, "An Appeal to the Young," in which he, as always, mixed politics and science: "It is now no longer a question of accumulating scientific truths and discoveries," Peter preached to the young reader of the newspaper that published his "Appeal":

> "We need above everything to spread the truths already mastered by science, to make them part of our daily life, to render them common property. We have to order things so that all, so that the mass of mankind, may be capable of understanding and applying them; we have to make science no longer a luxury but the foundation

of every man's life. This is what justice demands. I go further: I say that the interests of science itself lie in the same direction. Science only makes real progress when a new truth finds a soil already prepared to receive it."[5]

Kropotkin found much he could do with the Jura Federation. Writing pamphlets, "working out the practical and theoretic aspects of anarchist socialism,"[6] meeting with workers, attending rallies and protests, and co-founding—on an initial budget of the equivalent of four dollars—the revolutionary, fortnightly published newspaper, *Le Révolté*, consumed Peter's days and nights. In the midst of all that, though, Kropotkin fell in love, and in 1879 he married Sophie Ananiev, a Russian Jewish refugee living in Switzerland.

Though he occasionally traveled around Europe for short periods, including a trip to an Anarchist Congress in London, Peter's primary residence was in Switzerland until 1881. He was safe in his new country, until Czar Alexander II was assassinated in March, 1861 by Nikolai Rysakov, a member of the "People's Will"[7] movement. Alexander III, paranoid that the entire Czarist system might crumble, unleashed the military, the police and the officers of a newly minted "protection" agency called the Okhrana, to suppress any future acts of rebellion. Though Kropotkin was not associated with the People's Will group, and had no role in the assassination of Alexander II, the Swiss Government, under intense pressure from Alexander III to purge itself of Russian anarchists and agitators, expelled Peter and his wife.

The Kropotkins settled back in England, and spent a year in London, but soon felt the dearth of anarchists and socialists there, and packed their bags and their meager belongings and moved to Nice, France, where, as the result of an economic downturn, as well as bombings by local anarchist groups, unrest was at a high. With things at a political pitch, and aware of his ties to anarchists, the French police kept a sharp eye on Peter, even though Kropotkin, to his own dismay, had almost no ties to French anarchists. The Russian government also continued to monitor the ex-Prince: "Russian spies," he noted," began to parade again in conspicuous numbers in our small town."[8]

In December 1882, as part of a sweep that landed 65 anarchists in jail, Kropotkin was again arrested, this time for being a member of the International Workingmen's Association, an anarcho-socialist organization illegal in France. Peter knew his prospects were dim, as "a police court is always sure to pronounce the sentences which are wanted by the government."[9]

While in custody, awaiting trial, Kropotkin was offered a ticket to freedom, but he refused. A mysterious British friend had heard of Kropotkin's demise, and sent a messenger to France "with a considerable sum of money for obtaining my {Kropotkin's} release."[10] All that was required in return was for Kropotkin to "leave France immediately" for his own safety. Kropotkin, though touched at the magnanimous gesture, declined, as he felt that he needed to remain with the others arrested in the sweep.

Though the prosecutors had virtually no evidence in support of their case, in January 1883, Peter was found guilty and sentenced to five years in the Clairvaux prison. Kropotkin was unfazed—he had been through all this before in Peter and Paul, and was determined to make what he could of the situation. He was provided with some basics, including pen and paper, and used the time and the resources to write articles for *The Encyclopedia Britannica.*

Kropotkin still had many friends on the outside. The British guardian angel who offered to provide money for Peter's release was not the only foreigner that would play a part in this Kropotkin prison tale. Victor Hugo presented the French government with a petition to release Peter. The signatories of the petition included a group of British Members of Parliament, dozens of professors, leaders of the British Museum, editors of the *Encyclopedia Brittanica,* and a suite of newspaper editors.[11] Not everyone, though, was willing to affix their names to the cause—Thomas Henry Huxley, arguably the most well-known scientist in England, adamantly refused to sign this document. Kropotkin's political views rubbed Huxley the wrong way, and though Kropotkin did not yet know it, Huxley would soon become his *bête noire.*

By 1886, the French government, in response to continued pleas from the international community, released Kropotkin, who quickly returned to England, where a new socialist and anarchist movement was at last taking hold[12]— a prospect that frightened many, some of whom described anarchists as "a malignant fungoid growth...on the body politic...the very dregs of the population, the riff-raff of rascaldom, professional thieves [and] bullies."[13] Kropotkin was seen as something of the exception to this rule, as many in Britain were fond of his ideas on mutual aid and enamored with the audacity of his escape from Peter and Paul. So, while he did not have to worry about harassment from the British police (who actually used Kropotkin's essays on prisons in their own penal system), Peter was justifiably paranoid that the Russian secret police were still watching him. He continued, however, to preach his ideas on everything from mutual aid to anarchy and socialism. In fact, since he renounced his family's wealth as ill-gotten gains, Kropotkin's sole source of income was derived from his contributions to such papers and magazines as *The Times*, *Atlantic Monthly* and *Nature* where he was preaching these ideas.

In February 1888, Kropotkin opened the February issue of the popular Victorian magazine, *Nineteenth Century*, and was aghast at what he saw. Thomas Henry Huxley—the very man who had refused to sign the petition demanding Peter's release from French Prison—had written a long, scathing article entitled "The struggle for existence: a programme" in which he argued that nature was a dog-eat-dog bloodbath, not a Mecca of mutual aid. Kropotkin's response to what he called Huxley's "atrocious article"[14] would compel him to formalize his ideas on mutual aid, and eventually lead Peter to write his best-known book, *Mutual Aid: A Factor of*

Evolution. Kropotkin's parries to Huxley's atrocious article would also make him an international celebrity.

Thomas Henry Huxley, the youngest of six children, was born on May 4, 1825. As a child, he roamed "footloose in the silk-weaving city of Coventry."[15] "We boys," he later wrote in a short autobiographical essay, "were average lads, with much the same inherent capacity for good and evil as others; but the people who were set over us cared about as much for intellectual and moral welfare as if they were baby farmers. We were left to the operation of the struggle for existence among ourselves."[16]

From a young age, Huxley was enchanted with the natural sciences, and read whatever he could get his hands on, including Hutton's *Theory of the Earth* which he relished by candlelight before bed. It was men like Hutton that led Thomas to spend his evenings "speculating on the causes of colours at sunset" and the "nature of the soul and the difference between it and matter."[17]

Young Huxley, though, was not especially interested in "medicine as the art of healing," but was fascinated with "the mechanical engineering of living machines."[18] After two apprenticeships that included time in "a dissecting room {with} a naked cadaver, a cold body and a dead brain," that caused him to begin "questioning the meaning of evil and death," Thomas entered, and eventually completed, medical school. When he graduated in the spring of 1846, he found himself in debt, and to pay it off, he accepted a position as surgeon on the *HMS Rattlesnake*.

Like Kropotkin's journeys through Siberia, Huxley's time on the *Rattlesnake* was arduous, but taught him much about the art of survival: "it was good for me to live under sharp discipline; to be down on the realities of existence by living on bare necessities... cocoa and weevilly biscuit the sole prospect for breakfast."[19] More to the point, Huxley, like Darwin, had used his position on a ship to not only perform his duty (ship surgeon for the former, "companion to the Captain" for the latter), but also to gather immense amounts of information on the natural history of the places his ship visited. This work earned him a coveted Fellowship to the Royal Society at the tender age of twenty-five.[20]

Huxley began to come into his own in the 1860s, after engaging in a famous debate with Bishop Samuel "Soapy Sam" Wilberforce at the 1860 meeting of the British Association for the Advancement of Science.[21] He eviscerated Wilberforce while defending Darwin's ideas on evolution, and quickly became recognized as one of the leading thinkers of his day, writing and speaking on evolution, paleontology, anatomy, physiology, science education, religion and almost every subject that might be of interest to people of his age. Darwin, often too ill to speak in public about evolution, relied on Huxley to defend his ideas to the people, the media and other scientists; and Huxley was happy to do so, the only other option being "to let the devil have his own way."[22] Huxley's eclectic interests were given new fields in which to romp in 1864, when he initiated the X-Club—a group that would be a major force behind the course of Victorian science for almost the next thirty years.[23]

In 1888, Huxley published the essay that so infuriated Kropotkin. "The struggle for existence: a programme" was written at the same time that Thomas Henry was composing a much belated obituary for his friend and colleague Charles Darwin. By publishing it in *Nineteenth Century,* Huxley was getting a large audience, but an educated one: though it had a respectable circulation, *Nineteenth Century* was no British tabloid, with its contributors including Alfred Lord Tennyson, Beatrix Potter, Baron Rothschild and Prime Minister Gladstone.

In his article, Huxley wasted no time laying out his thesis about the natural world: "From the point of view of the moralist, the animal world is on about the same level as the gladiator's show. The creatures are fairly well treated, and set to fight; whereby the strongest, the swiftest and the cunningest live to fight another day. The spectator has no need to turn his thumb down, as no quarter is given ... ." The same could be said of ancient man, in his "savage" state: "the weakest and the stupidest went to the wall, while the toughest and the shrewdest, those who were best fitted to cope with their circumstances, but not the best in any other way, survived. Life was a continuous free fight, and beyond the limited and temporary relations of the family, the Hobbesian war of each against all was the normal state of existence."[24]

This became known as the "gladiator essay," and Huxley's transition from describing the forces shaping animal life, to outlining the human political condition, was seamless. Huxley easily moved from "The creatures are fairly well treated, and set to fight" in describing nature to "the effort of ethical man to work towards a moral end ... has hardly modified, the deep-seated organic impulses which impel the natural man to follow his nonmoral course" in describing the human condition. "Cosmic nature [*evolution*], Huxley would later say, "is no school of virtue, but the headquarters of the enemy of ethical nature."

Huxley steeped the human condition in the struggle for existence. He wasn't happy about having to describe mankind in those terms, but to be intellectually honest, he felt obliged to do so. He felt that ethics needed to be divorced from the evolutionary process. A deer running from a wolf may strike an emotional chord, even producing awe in the spectator at the grace of the hunted, but for Huxley, the wolf was equally skilled, and equally deserving of our admiration for doing what it does so well—all of which is to say that Nature is not immoral, just amoral.

For Huxley, the struggle for existence was enormous, and "no fiddle-faddling with the distribution of wealth will deliver society from the tendency to be destroyed by the reproduction within itself."[25] Huxley believed that almost nothing could ultimately stop this force, but he was not without hope; at least hope that we could fend off our self-destructive tendencies for some significant period of time. Indeed, the second half of Huxley's gladiator essay was devoted to outlining in very broad terms what English society might do, at least temporarily, to fend off the amoral underpinning of the evolutionary process.

For Huxley, human society was a good thing—he thought virtually everything decent about human life in 1888 was the result of creating societal rules and norms—but was always being eroded by the blind force of amoral evolutionary change. And he was under no impression that what he proposed—curtailing population growth, increasing productivity, creating better sanitation, and elevating science education—was some sort of panacea; Huxley's ideas were meant more as a stopgap measure than as an ultimate cure.[26, 27]

Thomas Henry Huxley wanted all to "understand, once and for all, that the ethical progress of society depends, not on imitating the cosmic process [evolution], still less in running away from it, but in combating it."[28] Man needed to revolt against an amoral nature, not return to it.

For Kropotkin, nothing could be further from the truth—we need not run from anything concerning our evolutionary, or for that matter, even our more contemporary roots, for all of these bask in the light of mutual aid. His five years in Siberia convinced Kropotkin that Huxley, brilliant as Kropotkin knew he was, was mightily misguided when it came to evolution and society—be it animal or human in nature. Peter was irate that Britain's leading scientist would use his soapbox to publish a piece like the gladiator essay. Fortunately, Kropotkin was friends with James Knowles, the editor of *Nineteenth Century*, and requested the opportunity to write a rebuttal.

Knowles granted Peter's request "with fullest sympathies,"[29] assuming what he would get was a brief critique of the Huxleyian depiction of nature.[30] Instead, Peter gave him a full-length essay called "Mutual Aid Among Animals." Indeed, this piece turned out to be the first of a series of "Mutual Aid in ..." essays (Mutual Aid in Savages, Mutual Aid in Barbarians) that would appear in

Nineteenth Century, but for Kropotkin it was arguably the most important.

If Siberia taught Kropotkin one thing, it was that "If we resort to an indirect test and ask Nature: 'who are the fittest', those that are continually at war with each other or those who support one another?, we at once see those animals which acquire mutual aid are undoubtedly the fittest."[31] But Peter understood that examples from Siberia would go only so far in making the generalized, sweeping claims he intended to make regarding the power of mutual aid in animals. He would need to show that mutual aid was a factor of animal life everywhere, not just on the frozen plains of Siberia.

"It was necessary," Kropotkin told his readers, "to indicate the overwhelming importance which sociable habits {mutual aid} play in Nature and in the progressive evolution of both the animal species and human beings." All of this was necessary, because, as Peter's article "Mutual Aid in Animals" spelled out to the reader, "the numberless followers of Darwin reduced the notion of struggle for existence to its narrowest limits. They came to conceive the animal world as a world of perpetual struggle among half-starved individuals, thirsting for one another's blood. They made modern literature resound with the war-cry of woe to the vanquished, as if it were the last word of modern biology."[32] Utter nonsense, Peter argued. Yes, it was true that competition was real, but it was a relatively weak force, and what Nature really shows at every turn is that "sociability is as much a law of nature as mutual struggle."[33] And for Peter, "sociability and intelligence always go hand in hand."[34]

Kropotkin's examples of mutual aid in animals include detailed descriptions, as well as vague generalities.[35] As examples of the latter, Peter mentions that falcons, which possess "an almost ideal organization for robbery" are on the decline, while the duck "practices mutual support, and it almost invades the earth."[36] Mutual aid at this generalized level was apparent everywhere—even in pond scum: "mutual aid is met with even amidst the lowest animals, and we must be prepared to learn some day, from the students of microscopical pond-life, facts of unconscious mutual support, even from the life of microorganisms."[37]

Kropotkin's examples are not limited to such vague statements about animal life. Take the case of the burying beetle, who lays its eggs in decomposing rodent carcasses: "As a rule," Peter writes, these creatures, "live an isolated life, but when one of them has discovered the corpse of a mouse or of a bird, which it hardly could manage to bury itself, it calls four, six, or ten other beetles to perform the operation with united efforts; if necessary, they transport the corpse to a suitable soft ground; and they bury it in a very considerate way, without quarrelling as to which of them will enjoy the privilege of laying its eggs in the buried corpse."[38]

With their complex societies, the social insects—the ants, bees, and wasps—were special favorites of Peter, and he regaled in communicating tales of mutual aid in these creatures. Consider what happens when an ant comes back to its nest hungry, and solicits food from one of its nestmates. The solicitor and its partner, Peter tells his reader, "exchange a few movements with the antennae, and if one of them is hungry or thirsty, and especially if the other has its crop full... it immediately asks for food. The individual thus requested never refuses." This example was but one of many that Kropotkin wrote of ants, but it held a special place in the pantheon of mutual aid, for here behavior seems to shape the very anatomy of the ant. "Regurgitating food for other ants is so prominent a feature in the life of ants," Peter writes, "that Forel considers the digestive tube of the ants as consisting of two different parts, one of which, the posterior, is for the special use of the individual, and the other, the anterior part, is chiefly for the use of the community."[39] Even mutual aid's chief proponent could not have asked for more from an animal system.

But it wasn't just that mutual aid was *present* in ants that so fascinated Peter. It was the way that ant cheaters, who refused to dispense aid, were treated, that convinced Kropotkin of the power of mutual aid. "If an ant which has its crop full has been selfish enough to refuse feeding a comrade, it will be treated as an enemy, or even worse. If the refusal has been made while its kinsfolk were fighting with some other species, they will fall back upon the greedy individual with greater vehemence than even upon the enemies themselves." Punishment enforced mutual aid, even in ants. "If we knew no other facts from animal life than what we know about the ants and the termites," Kropotkin proclaims, "we already might safely conclude that mutual aid…and individual initiative…are two factors infinitely more important than mutual struggle in the evolution of the animal kingdom."[40]

Enamored as he was with insects, Kropotkin also describes acts of mutual aid in other creatures. Eagles, for example, employed mutual aid to hunt their prey and to distribute the food amongst themselves. Peter writes of a naturalist who had seen "an eagle belonging to an altogether gregarious species (the white-tailed eagle, *Haliactos albicilla*), rising high in the air…when at once its piercing voice was heard. Its cry was soon answered by another eagle which approached it, and was followed by a third, a fourth, and so on, till nine or ten eagles came together and soon disappeared." Later, the same naturalist "went to the place whereto he saw the eagles flying…and discovered that they had gathered around the corpse of a horse. The old ones, which, as a rule, begin the meal first—such are their rules of propriety—already were sitting upon the haystacks of the neighbourhood and kept watch, while the younger ones were continuing the meal."[41] Such avian mutual aid during hunts was not restricted to eagles. "It would be quite impossible to enumerate here the various hunting associations of birds," Kropotkin writes, "but the fishing associations of the pelicans are certainly worthy of notice for the remarkable order and intelligence displayed by these clumsy birds. They always go fishing in numerous bands, and after having chosen an appropriate bay, they form a wide half-circle in face of the shore, and narrow it by paddling towards the shore, catching all fish that happen to be enclosed in the circle."[42] So ubiquitous was such mutual aid in avian foraging that Kropotkin spoke of it as an "established fact."[43]

Mammals too hunted prey in a cooperative fashion. But mutual aid could also be used as a defensive, as well as an offensive, tool. Horses, for example, lived in groups and cooperatively fended off their enemies. "When a beast of prey approaches them, several studs unite at once; they repulse the beast and sometimes chase it: and neither the wolf nor the bear, not even the lion, can capture a horse or even a zebra as long as they are not detached from the herd...Union is their chief arm in the struggle for life."[44]

And, of course, mutual aid permeated the primates, which Peter described as "sociable in the highest degree."[45] "Several species," Kropotkin wrote admiringly, "display the greatest solicitude for their wounded, and do not abandon a wounded comrade during a retreat till they have ascertained that it is dead and that they are helpless to restore it to life."[46]

Though Kropotkin intentionally focused on mutual aid in animals in their natural environment, he was not above relaying personal anecdotes of mutual aid in more artificial settings. Once, when Peter was in the Brighton Aquarium, he became transfixed on a crab that had flipped over on its shell, and become immobilized. To Kropotkin's utter delight, "Its comrades came to the rescue, and for one hour's time I watched how they endeavoured to help their fellow-prisoner. They came two at once, pushed their friend from beneath, and after strenuous efforts succeeded in lifting it upright; but then the iron bar would prevent them from achieving the work of rescue, and the crab would again heavily fall upon its back." But this aquatic demonstration of mutual aid was hardly over. "After many attempts," Peter writes, "one of the helpers would go in the depth of the tank and bring two other crabs, which would begin with fresh forces the same pushing and lifting of their helpless comrade."[47]

In the last few words of Kropotkin's exposition on mutual aid in animals, we see Peter the poet summarizing Peter the natural historian's conclusions on mutual aid: "Don't compete!—competition is always injurious to the species, and you have plenty of resources to avoid it! That is the tendency of nature, not always realized in full, but always present. That is the watchword which comes to us from the bush, the forest, the river, the ocean. Therefore combine—practise mutual aid! That is the surest means for giving to each and to all the greatest safety, the best guarantee of existence and progress, bodily, intellectual, and moral. That is what Nature teaches us; and that is what all those animals which have attained the highest position in their respective classes have done."[48]

While Kropotkin was far and away the most vociferous and well-known spokesman for mutual aid in animals, he was not alone, nor was he the originator of the idea.[49] The founder of the mutual aid school of Russian evolutionary thinking was Karl Fedorovich Kessler.[50] Originally trained as a mathematician, Kessler eventually became a world-class naturalist, and the author of a massive five-volume series on the birds, mammals and fish of the Kiev region of Russia. As with Kropotkin, wherever Kessler went to study natural history, he witnessed what he took to be animals and humans employing mutual aid combating a fierce environment. In December 1879, Kessler gave a speech entitled "On the Law of Mutual Aid" to the St. Petersberg Society of Naturalists. For Kessler, Darwin's "individual vs. individual" struggle was secondary to the mutual aid that organisms so readily display. When Kessler died in 1881, Kropotkin took over as leader of the Russian mutual aid camp.

Kropotkin, and the Russian school in which he was trained, admired much about Darwin. They believed it was not Darwin, but his followers, who perverted his ideas into a natural bloodbath of sorts. Led by Kropotkin, this group argued that the "phony Darwinists,"[51] such as Huxley, were omnipresent, and that such "vulgarizers of the teachings of Darwin have succeeded in persuading men that the last word of science was a pitiless individual struggle for life."

For Darwin, the struggle for existence *sometimes* meant a literal struggle: "Two canine individuals in a time of dearth may be truly said to struggle with each other which shall get food and life."[52] Darwin summed up his ideas on this succinctly when noting: "As more individuals are produced than can possibly survive there must be in every case a struggle for existence...It is the doctrine of Malthus applied with manifold force to the whole animal and vegetable kingdom."[53] These ideas were ingrained not only in Darwin's views, but in those of his colleagues, such as Huxley, who in 1873, a good fifteen years before the gladiator essay, left no room for doubt about the relationship between Malthus and natural selection: "It {natural selection} is indeed simply the law of Malthus exemplified... ."[54]

But Darwin's struggle could be more metaphorical—a struggle of individuals against environment rather than individuals against individuals. Indeed, in *The Origin of Species,* the sentence following Darwin's description of two canines fighting for food, goes on to say, "But a plant on the edge of the desert is said to struggle for life against drought... ." A more Russian example might be a herd of mammals struggling against the freezing cold of Siberia by staying close to one another and keeping warm.

Direct competition between individuals, like the canines in Darwin's first example, came as a result of overpopulation in the face of limited resources. This could lead to the nasty world depicted by Huxley. But, *under*population, Kropotkin and the Russian school believed, would often lead to animals grouping together to exhibit mutual aid in the face of environmental challenges. "Paucity of life, under-population—not over-population," Kropotkin noted, caused "serious doubts...as to the reality of that fearful competition for food and life within each species, which was an article of faith with most Darwinists, and, consequently, as to the dominant part which this sort of competition was supposed to play in the evolution of new species."[55]

Darwin's "vulgarizers," Kropotkin argued, focused solely on overpopulation and hence competition. But Peter's experiences in Siberia suggested that underpopulation better represented nature, which made mutual aid the default result of the evolutionary process. The real question was not whether Darwin was "right"—"Life is struggle," Kropotkin wrote, "and in that struggle the fittest survive. But the answers to the questions, 'By which arms is this struggle chiefly carried on?' and 'Who are the fittest in the struggle?' will widely differ according to the importance given to the two different aspects of the struggle: the direct one, for food and safety among separate individuals, and the struggle which Darwin described as 'metaphorical'–the struggle, very often collective, against adverse circumstances."[56]

Underpopulation, and the struggle of the individual versus Nature that arose from it, was the key to distinguishing between the two forms of conflict which Peter described. Kropotkin railed against "the supposed pressure of population on the means of subsistence," which he saw as "...mere fallacy, repeated, like many fallacies, without even taking the trouble of submitting it to a moment's criticism."[57]

The Russian intellectual community, in most instances, rejected Malthusian overpopulation, not only because the world they lived in did not match that set out in Malthus' work, but because the work reeked of British individualism: As one of the leading Russian scientists of the day noted, "The essential, dominating characteristic of the English national character is love of independence, the all-sided development of the personality, and individualism; which manifests itself in a struggle against all obstacles presented by external nature and other people. Struggle, free competition, is the life of the Englishman; he accepts it with all its consequences, demands it as his right, tolerates no limits upon it."[58]

The Russians faced a dilemma. They thought highly of Darwin and his ideas on evolution. Indeed, Darwin's theory of evolution per se received a smoother reception in Russia than in England or America. At the same time, for most Russians, Darwin's tie to Malthusian overpopulation was unacceptable, thus creating a conundrum: how to accept the former, but reject the latter?

Within the Russian scientific community, the solution to this dilemma was to admit that the tie existed, but to claim that it was far overblown; that Darwin's ideas worked perfectly well without the Malthusian component. And when Malthus was removed from Darwin, the Russian school got mutual aid—and lots of it. This love-hate relationship with Darwin was nicely encapsulated by V.V. Dokuchaev, a Russian soil scientist, who both praised and criticized Darwin in the same breath, and at the same time mixed religion, science and politics into one hodgepodge: "The great Darwin, to whom contemporary science is indebted for perhaps 9/10 of its present scope, thought that the world was governed by the Old Testament law: an eye for an eye, a tooth for a tooth. This was a big mistake, a great confusion. One should not blame Darwin for this error...But Darwin, thank God, turns out to have been incorrect. Alongside the cruel Old Testament law of constant struggle we now clearly see the law of cooperation, of love... ."[59]

The Russian school believed that the great law of Darwin, of course, had implications for mutual aid in humans, as well as animals. Peter Kropotkin's next mission was to demonstrate this to Huxley—and to everyone else.

≡ CHAPTER 5 ≡

STRANGE BEDFELLOWS

"The importance of mutual aid in the evolution of the animal world and human history may be taken, I believe, as a positively established scientific truth, free of any hypothetical admission."

Peter Kropotkin[1]

For Peter, mutual aid was natural, whether in bird or peasant. But Kropotkin needed to convince others of this. He needed a theory, above and beyond an evolutionary one, that would explain why foraging birds helped each other and why peasants welcomed strangers into their homes. He needed what today we might call a "real time" theory that centered on the immediate cause that led to these acts of kindness.

For this, Kropotkin turned to the ideas of economist Adam Smith, who was already regarded as one of the founders of the field of economics. Smith's 1776 book, *An Inquiry into the Nature and Causes of the Wealth of Nations*, in which he developed his theory of capitalism, was required reading for the intelligentsia of Kropotkin's day. Of course, for anarcho-socialists like Kropotkin, *The Wealth of Nations* espoused precisely the wrong economic system and was reviled as a dangerous weapon that was used to suppress the masses. But Peter had a soft spot for the younger Adam Smith, who, in 1759, had published *The Theory of Moral Sentiments,* which Kropotkin claimed was "far superior to the work of his {Smith's} old age upon political economy... {Smith} sought the explanation of morality in a physical fact of human nature."[2] It was in Adam Smith's "morality in a physical fact of nature" that Peter found a causal theory of mutual aid in both humans and animals.

In *The Theory of Moral Sentiments*, Adam Smith argued that because humans want to minimize their own pain, and because we are naturally empathetic, we sometimes act to minimize the pain of others, in order to minimize our own empathetically induced pain. "How selfish soever man may be supposed," Smith wrote, "there are evidently some principles in his nature, which interest him in the fortune of others...of this kind is pity or compassion, the emotion which we feel for the misery of others, when we either see it, or are made to conceive it in a very lively manner. The greatest ruffian, the most hardened violator of the laws of society, is not altogether without it."[3] When we help those in need, we minimize our own vicarious pain, Smith wrote, "without any consideration of their tendency to those beneficent ends which the great Director of nature intended to produce by them."[4]

While Peter was naturally disinclined to ally himself with the founder of capitalism, he saw the greater good in doing so, and in moving beyond evolutionary theories alone. Kropotkin found Smith's ideas on empathy and mutual aid powerful and elegantly simple. "You see a man beat a child," Peter wrote, and "you know that the beaten child suffers. Your imagination causes you yourself to suffer the pain inflicted upon the child; or perhaps its tears, its little suffering face tells you. And if you are not a coward, you rush at the brute who is beating it and rescue it from him. This example by itself explains almost all the moral sentiments."[5] In Kropotkin's view, the only shortcoming in *The Theory of Moral Sentiments* was that Smith did not take his theory of empathy far enough.

All of Adam Smith's examples of empathy and mutual aid were human examples. But, why, Kropotkin wondered, should so powerful an explanation for mutual aid be limited to our own species? "Adam Smith's only mistake," Peter wrote, "was not to have understood that this same feeling of sympathy in its habitual stage exists among animals as well as among men."[6] The same mechanism—instinctive empathy—explained why animals and humans came to each other's aid.

Empathy was the driver of animal solidarity, and from solidarity followed evolutionary success, because it led to confidence that aid would be dispensed when it was required: "without mutual confidence," Kropotkin wrote, "no struggle is possible; there is no courage, no initiative…! Defeat is certain."[7] From empathy to solidarity to mutual aid. This was a powerful formula for Kropotkin; powerful enough that it led him to wild, if poetic, speculation:

> "… let us imagine this feeling of solidarity acting during the millions of ages which have succeeded one another since the first beginnings of animal life appeared upon the globe. Let us imagine how this feeling little by little became a habit, and was transmitted by heredity from the simplest microscopic organism to its descendants—insects, birds, reptiles, mammals, man—and we shall comprehend the origin of the moral sentiment, which is a necessity to the animal like food or the organ for digesting it."

Microscopic solidarity aside, Adam Smith's ideas on empathy and mutual aid in humans had provided Kropotkin with a powerful theoretical construct that he would apply many times to animals and humans alike, especially in his book *Mutual Aid: A Factor in Evolution*.

Peter was now armed with theory—both evolutionary and real time—and his own confidence that mutual aid was the predominant force driving human evolution. But he knew of the resistance he faced among an audience conditioned to see human history as one of ceaseless competition and endless wars. So, after two chapters reviewing cooperation among animals in *Mutual Aid: A Factor of Evolution,* it was time for Kropotkin "to cast a glance upon the part played by the same agencies in the evolution of mankind."[8] Much more than a glance, *Mutual Aid* has six full chapters on cooperation in humans: *Mutual Aid Among Savages*, *Mutual Aid Among the Barbarians*, *Mutual Aid in the Medieval City (I & II)* and *Mutual Aid Amongst Ourselves (I & II)*.

Kropotkin would have to contend with a readership that embraced Thomas Hobbes, who, in *Leviathan,* had described "the life of man {as} solitary, poor, nasty, brutish, and short."[9] Indeed, Thomas Henry Huxley had singled out "the Hobbesian war of each against all" as "the normal state of existence"[10] for all creatures, man included. Peter thought such a depiction of human nature was not only "utterly indefensible, improbable and unphilosophical,"[11] but unsupported by an in-depth analysis of human history.

Kropotkin began his overview of human mutual aid by arguing that Hobbes, Malthus, Huxley and many other political philosophers, scientists and historians had erred in assuming that the first human societies revolved around the family unit. "Science has established beyond any doubt that mankind did not begin its life in the shape of small isolated families," Kropotkin wrote, "…as far as we can go back in the palaeo-ethnology of mankind, we find men living in societies—in tribes similar to those of the highest mammals."[12]

In a tribal setting, mutual aid could be dispensed regardless of blood ties, making it a powerful, omnipresent, evolutionary force.

Kropotkin invoked archeological evidence noting that stone tools and pottery were often found in such vast quantities—"heaps from five to ten feet thick, from 100 to 200 feet wide, and 1,000 feet or more in length" at one site in Denmark[13]—and could only be the remnants of a tribal, not familial, social structure.

Kropotkin's exploration of mutual aid started with the anthropological literature, which to him suggested that the Bushmen of Africa never quarreled, divided their food equally between all group members, and were quick to aid their brothers in distress. The Hottentots, though described as filthy animals by many ethnologists of the day, displayed the same forms of mutual aid as the Bushmen, and, as Peter noted with no small degree of satisfaction: "If anything is given to a Hottentot, he at once divides it among all present—a habit which, as is known, so much struck Darwin among the Fuegians."[14]

Long before Margaret Mead,[15] Kropotkin wrote of "the harmony which prevails in the villages of the Polynesian inhabitants of the Pacific Islands" who practiced what he described as an equal distribution of possessions without thought to individual gain.[16] And so too did the Eskimos. When an Eskimo became too rich, Peter noted, he would hold a giant feast in which he divided up all his possessions among those in his tribe. Such acts of equality minimized jealousy and competition and made mutual aid the driving force of Eskimo culture.

Cross-cultural studies in anthropology, archeology and ethnology convinced Kropotkin that "the high development of tribal solidarity and the good feelings with which primitive folk are animated towards each other, could be illustrated by any amount of reliable testimony."[17] His unbounded confidence in mutual aid as *the* factor driving evolution led Kropotkin to suggest that even infanticide and parricide (the killing of the old and feeble) were acts of mutual aid. For a society to flourish and have the means to provide for all its members, infanticide and parricide became acts of altruism and courage, necessary to avoid overpopulation and sustain tribal existence. "West European men of science," Kropotkin wrote, "when coming across these facts, are absolutely unable to understand them,"[18] but once history was seen through the lens of mutual aid, the role of infanticide and parricide were more easily understood.

Among the goals of *Mutual Aid* was a corrective to the myopic view of human history promulgated by Hobbes, Malthus, Huxley and others, by Kropotkin's use of "a minute analysis of thousands of small facts and faint indications accidentally preserved in the relics of the past."[19] And though Peter did find mutual aid in every society that he studied, he was troubled by what he called the "double conception of morality." Mutual aid was omnipresent among members of a group, but there seemed to be a different set of rules in play between individuals from different groups. Inter-tribal interactions often resulted in aggression, and sometimes war.

Kropotkin found reconciliation in his unbounded belief in the power of mutual aid to guide animal and human affairs, but this led him down a tortured path. Intertribal aggression, he argued, could still be thought of as mutual aid, because attacking outsiders was the ultimate act of mutual aid to one's own group. This was certainly not the type of mutual aid Kropotkin wanted humans to display, but it afforded him an explanation that preserved mutual aid and its evolutionary roots. For Peter, what this implied was that tribal identity, and equally important, tribal territory, had to be expanded, reducing the number of "others," and increasing the number of fellow groupmates with whom mutual aid could be exchanged.

Kropotkin's sortie into human history suggested that such an expansion of tribal identity and territory had occurred after the Barbarians swept across the plains of Europe and Asia and eventually caused the fall of the Roman Empire.[20] The next stage of mutual aid was then set in motion.

When the Barbarian invasions were over, and sufficient time had passed for the establishment of new societies, mutual aid was channeled not through the tribe, as it had been among savages, but in the new village communities that had arisen. While some private wealth could be amassed by individuals in villages, the defining feature of the village community was the commons area. A classic example of mutual aid, part of the commons area was assigned to communal gardens, and part was set aside as an area from which all livestock from the village could graze freely. Tribunals were then elected to handle any quarrels that might emerge over the use of the commons.[21] It was the commons, the tribunals that oversaw them, and the benefits that they produced via mutual aid, which became, according to Peter, "the chief arm" in the "hard struggle against a hostile nature."[22]

"There being no authority in a village community to impose a decision," Kropotkin wrote, "this system has been practiced by mankind wherever there have been village communities, and it is practiced still wherever they continue to exist, i.e. by several hundred million men all over the world."[23] Peter the anarchist could hardly have asked for more.

In these communal villages, Peter argued, selfish men eventually devised ways to profit from the mutual aid displayed by villagers, and slowly began to take power and run villages for their own benefit. Mutual aid declined as a result, only to emerge from the ashes in the "free cities" of the Medieval Period. These free cities, which were governed only from within, arose, Peter argued, when the suppressed masses united and rose up against the lords and clerical masters that had subjugated them for centuries. Kropotkin described the rise of such cites as "a natural growth in the full sense of the word"[24] and one that always involved the same process, namely:

> "The movement spread from spot to spot, involving every town on the surface of Europe...free cities had been called

into existence on the coasts of the Mediterranean, the North Sea, the Baltic, the Atlantic Ocean, down to the fjords of Scandinavia; at the feet of the Apennines, the Alps, the Black Forest...in the plains of Russia, Hungary, France and Spain. Everywhere the same revolt took place, with the same features, passing through the same phases, leading to the same results....their "co-jurations," their "fraternities," their "friendships," united in one common idea, and boldly marching towards a new life of mutual support and liberty. And they succeeded so well that in three or four hundred years they had changed the very face of Europe."[25]

The guilds of these free cities created a mutual bond as great as any Kropotkin had ever witnessed. The common marketplaces, the exquisite new buildings that were constructed, the joint defense against any attacks from the outside, were all examples of the mutual aid that flowed from the guilds of stonemasons, carpenters, merchants, painters, teachers, and musicians that coexisted within the walls of the medieval city. In such a society, where people lived peacefully with each other, they were free to think in new ways, to extend the limits of the human mind, and within these walls, Kropotkin argued, inductive science took hold.

In such a setting, there was always the possibility of selfish behavior, as it had occurred in the communal villages. To counter social parasitism, guilds established strict penalties on cheaters, and free cities established charters, built their fortifications, and created councils to negotiate internal disputes. Without any central government or dictators or lords, Peter argued, mutual aid governed these cities.

The ideas of justice, fairness and "right" permeated Peter's descriptions of the medieval city: even bread must be "baked in justice." [26] Laws established in free cities prohibited laborers from being forced to work more than forty hours a week. Children, too, were protected from harm: "although school meals did not exist, probably because no children went hungry to school," Peter wrote, "a distribution of bath-money to the children whose parents found difficulty in providing it was habitual in several places."[27] Such free cities were like living, breathing organisms for Kropotkin: they maintained their homeostasis through mutual aid amongst their components.

These cities presented Peter with the same moral duality that plagued his thoughts about earlier stages of human history. Mutual aid flourished within groups, but was occasionally protected by violence against outsiders: "In reality," Peter admits, "the medieval city was a fortified oasis amidst a country plunged into feudal submission, and it had to make room for itself by the force of its arms." Some free cities had to defend themselves from attackers; others had to form alliances to ward off larger dangers. Imperialist historians, as Peter liked to call them, suggested that free cities (and free city alliances) had initiated, and even reveled in, such battles, but Kropotkin was certain that the bulk of the evidence suggested otherwise—that, as always, the taking up of arms against outsiders was used only as a last resort to secure mutual aid within the threatened society.

Kropotkin saw medieval free cities as a grand experiment, the results of which enlightened men for centuries to come: "a close union for mutual aid and support, for consumption and production, and for social life altogether, without imposing upon men the fetters of the State, but giving full liberty of expression to the creative genius of each separate group of individuals in art, crafts, science, commerce, and political organization."[28]

Medieval free cities flourished for hundreds of years, and then they gradually disappeared. But in their disintegration, Peter found valuable lessons. On certain historical occasions, splinter groups, sometimes, but not always, revolving around a guild, began to form and work autonomously—mutual aid was dissolved from within. The birth of larger states across Europe and Asia also threatened the free city, both by physical force and by campaigning on the value of centralized governments.

Peter found vestigial traces of mutual aid, proof that it hadn't been crippled with the demise of the medieval free city. "It flows still even now," Kropotkin noted, "and it seeks its way to find out a new expression which would not be the State, nor the medieval city, nor the village community of the barbarians, nor the savage clan, but would proceed from all of them, and yet be superior to them in its wider and more deeply humane conceptions."[29] For Kropotkin, that "new expression" was anarcho-socialism.

Once he had brought his readers up to the anarcho-socialist movement, Kropotkin had completed his tour-de-force *history* of mutual aid in animals and humans. Next, Peter decided to show that the relatively rapid pace at which mutual aid appeared and spread was perfectly in line with biological expectations. Here, he had a problem. Kropotkin was a Darwinist, in the sense that he believed that natural selection drove evolutionary change. But Darwinian theory seemed to challenge the speed at which mutual aid operated, because natural selection was supposed to be a slow, gradual process that unfolded over eons. Yet, whenever animals or humans were placed in harsh new environments, Kropotkin's research had shown that mutual aid appeared rapidly. How could he reconcile this with the slow pace of natural selection?

Peter turned to the work of 18th century botanist and zoologist Jean-Baptiste Lamarck.[30] In his 1809 book, *Zoological Philosophy*,[31] published half a century before Darwin's *Origin of Species*, Lamarck proposed a theory for how evolutionary change might result from individuals adapting to their environments. He hypothesized that changes in the habits of an organism during its lifetime would be passed down to the next generation—an idea that has come to be known as the inheritance of acquired characteristics. For example, Lamarck observed that long legs appeared to be beneficial to shore birds who might sink in the sand otherwise. But how did long legs evolve?³² "One may perceive that the bird of the shore, which does not at all like to swim, and which however needs to draw near to the water to find its prey, will be continually exposed to sinking in the mud," Lamarck began. "Desiring to avoid immersing its body in the liquid [it] acquires the *habit* of stretching and elongating its legs. The result of this for the generations of these birds that continue to live in this manner is that the individuals will find themselves elevated as on stilts, on naked long legs."³³

This idea that changes in the habits of birds—stretching their legs in this case—could lead to the next generation of birds having longer legs is a radically different evolutionary process than Darwin would propose five decades later. For Darwin, those variants of a trait—longer legs, shorter legs and so on—that led to greater reproductive success would slowly increase over time, but the actions of an individual during its lifetime (the stretching of its legs) could not affect the variety of the trait that was passed on to the next generation. Both Darwin's and Lamarck's processes lead to organisms becoming better adapted to their environment, but they differ in the way that traits are transmitted across generations.³⁴ A Lamarckian process involving the inheritance of acquired characteristics can dramatically speed up the evolutionary process—the actions of individuals in a single generation can have a large effect on what traits and behaviors are present in the very next generation.³⁵

The implications of this speedier evolutionary process for mutual aid were profound, and Kropotkin wrote a series of articles, again in *Nineteenth Century,* on the power of Lamarck's inheritance of acquired characteristics.³⁶ "A synthesis of the views of Darwin and Lamarck," Peter told the reader of *Nineteenth Century*, provides "a naturalistic conception of the universe, the very foundations of human ethics."³⁷ Ethics, the Prince believed, were meaningless in the absence of mutual aid, and Lamarck's ideas provided the engine for mutual aid to spread with lightning speed. This should give pause, Peter argued, to the vulgarizers—Huxley, Hobbes and

Malthus in particular—who had argued so passionately that evolution by natural selection led to a merciless war of all against all.

Lamarckian inheritance provided the additional speed Kropotkin needed to reconcile evolutionary theory with his observations on how quickly mutual aid emerged in humans and animals,[38] and convinced Peter that he now had evolution's two most famous thinkers—Darwin and Lamarck—on his side.

≡ CHAPTER 6 ≋

WE ARE ALL BRETHREN

"whatever the wars they have fought, mere short-sighted egotism was at the bottom of them all."

Peter Kropotkin[1]

Kropotkin's mind was never at rest. He thought, wrote, and spoke about an astonishing array of topics: evolution, anarchism, socialism, communism, natural history, geology, the coming industrial revolution in the East, the French Revolution, the ideals and realities of Russian literature, ethics, and educating the next generation of revolutionaries.[2] As diverse as these topics may seem, for Peter, they were mere variations on a theme—mutual aid. He found that the implications of mutual aid were boundless, and he wanted others to be persuaded of the same.

Peter's attempts to find mutual aid undergirding every human action took him in some surprising directions. His essay entitled "What Geography Ought to Be" is a nice example of the unexpected twists and turns his audiences experienced. In 1899 this essay appeared in *Nineteenth Century*—the same magazine in which Kropotkin had published his articles on mutual aid. "What Geography Ought to Be" was a companion piece to a more technical report by the Royal Geographical Society (RGS) on the state of geographical education.[3] The RGS report, authored by John Scott Keltie, concluded that the state of geography in the British school system was dismal: it was taught poorly and rarely, and Britain trailed far behind its European counterparts. The RGS knew, all too well, that this report would never be read by the general public, and so, despite the fact that Keltie thought Kropotkin's political ties "seriously diminished the services which otherwise he might have rendered to Geography,"[4] he and the entire RGS took pleasure in the publication of Kropotkin's popular companion piece to their report.[5]

Part of Peter's essay addressed practical questions like how geography teachers should be trained, and how networks of young geographers from around the world could exchange rock samples they had collected.[6] But Kropotkin also took the occasion to critique teachers, at whose hands geography, despite its inherent excitement, had become "an arid and unmeaningful subject."[7]

For Kropotkin, geography, then considered the study of the earth's surface, was a subject with the potential to engage young minds, if presented in the right fashion. "Tales of hunting and fishing, of sea travels, of struggles against dangers," Peter told his reader, "...will utilize the child's imagination, not for stuffing it with superstition but for awakening the love of scientific studies."[8] These very same tales had cast a spell over Kropotkin at a young age, and could be used to teach children that struggles against nature were what drove mutual aid in humans. The desire to survive in a world full of dangers such as droughts, freezing temperatures, hurricanes and volcanic eruptions forced collaboration—they forced mutual aid. Natural selection, combined with the inheritance of acquired characteristics, Peter argued, favored acts of mutual aid and in so doing, transformed the human response to geographic dangers into a biologically inherited behavior.

Ever one to take an argument to the extreme, Peter proclaimed that geography taught students about the unifying power of mutual aid. The study of geography, he wrote, "teaches us, from our earliest childhood, that we are all brethren, whatever our nationality...that whatever the wars they have fought, mere shortsighted egotism was at the bottom of them all."[9] In his view, the borders of countries were merely vestiges of a barbaric past.

"What Geography Ought to Be" was just one salvo in Peter's arsenal of weapons showing that mutual aid had practical implications. Implications, for example, about crime and punishment. Along these lines, in December 1877, and then at much greater length in his book *In Russian and French Prisons*,[10] Kropotkin drew heavily on his time inspecting prisons in Siberia and his experiences in both the Peter and Paul and Clairvaux prisons as source material.

The challenge to a system governed by mutual aid was in transgression—in what to do with cheaters. If mutual aid was to flourish, then society would have to have some mechanism for dealing with rule breakers. But, incarceration, Peter believed, was not the ideal. The prison system failed in its claims of reformation and it was flawed at a deeper level—it stripped a person's abilities to contribute to mutual aid once beyond prison walls. "The prison," Kropotkin decreed, "kills all the qualities of a man which make him best adapted to community life."[11] It did this by cutting prisoners off from every aspect of the outside world—most especially for Peter, access to books. And it destroyed a respect for work in the pittance it paid to forced laborers. Such places, Peter said, were no more than "monuments of human hypocrisy and cowardice."[12]

Kropotkin argued that no reforms could remedy such a fatally flawed system; he called for the permanent closure of prisons. In Peter's vision of the ideal society, prisons would be anathema. "Antisocial acts," Kropotkin believed, "need not be feared in a society of equals...all of whom have acquired a healthy education and the habit of mutually aiding one another." Yes, Peter acknowledged, in the best of societies there would occasionally be a rule-breaker, but the "only practical corrective still will be fraternal treatment and moral support." [13]

Kropotkin's prolific writings portray a mind unable to contemplate any limits to mutual aid. As predictable as the themes of his writings had become, he was still occasionally led by his obsession down unexpected paths: one of which took him to the French Revolution. Peter saw the French Revolution as an exemplar of the strengths of man to forge a moral society.

In the late 1870s, Kropotkin entrenched himself in the British Library and delved deeply into its literature on the French Revolution.[14] He would have preferred access to the primary literature in the Bibliothèque Nationale de France, but Peter was persona non grata in Paris.[15] His forays into the British Library archives eventually culminated in his largest tome, *The Great French Revolution* (1909).[16] This detailed accounting of the revolution, told from the perspective of a working class peasant, served as an homage to mutual aid.

Peter argued that the French Revolution was actually composed of two different, but related, revolutions. Most significant was the economic revolution, the movement toward a communal economic system, where liberty and the good life reigned. The French Revolution—or, as Peter always referred to the event, the Great French Revolution—was, in Kropotkin's opinion, a peasant-based economic insurrection designed to gain rightful possession of the land.[17] It was also a political revolution, in which the monarchy was ousted by a series of alternative systems of governance. Kropotkin argued that critics who asked "was it worth it?" in the face of Napoleonic despotism missed the point, which was that things in France could never be the same. Though the revolution may have failed, once the people had tasted the fruits of mutual aid—even the bitter fruits—they would ultimately never be satisfied with anything less.

"A new France was born during those four years of revolution," Peter told his reader. "For the first time in centuries the peasant ate his fill {and} straightened his back,"[18] because the communal system in which the people were given back the land that was rightly theirs, led to a surge in productivity. With the freedom to work together—when mutual aid was allowed to flourish—the people had food and all that they needed to be très content.

Peter concluded *The Great French Revolution* with an ambitious attempt to unify history, politics and the science of mutual aid. Returning to his observations of animal and human cooperation in Siberia, Kropotkin reminded his readers that mutual aid was especially critical in harsh environments, and that, indeed, these harsh environments, combined with the Lamarckian inheritance of acquired characteristics, led to cooperation becoming a fundamental aspect of being human. The French Revolution was the quintessential example of this, leading to mutual aid amongst the people of France: "born out of the pressing necessities of those troubled years," Kropotkin wrote, " the communism of 1793, with its affirmation of the right of all to sustenance and to the land for its production, its denial of the right of any one to hold more land than he and his family could cultivate—this was the source and origin of all the present communist, anarchist, and socialist conceptions."[19]

The political consequences of mutual aid reached far beyond the French Revolution. Discussions of mutual aid, in one form or another, permeated much of Kropotkin's writings on politics, but were especially prominent in his book *The Conquest of Bread* (1892), his pamphlet *Modern Science And Anarchism* (1908), and his article, "The Scientific Bases of Anarchy (1887)," which appeared in *Nineteenth Century*.

The roots of anarchy—which Kropotkin defined as the "no-government system of socialism"[20] ran like rhizomes to mutual aid.

And though Peter certainly thought much about the philosophical implications of anarchy, the concept itself was not an abstract one, linked to the "studios of philosophers." Instead, anarchy and its underlying theory of mutual aid, found their home in the natural sciences. "The anarchist thinker does not resort to metaphysical conceptions like the natural rights {and} the duties of the State,"[21] Peter assured his readers. Instead, the theory of anarchy rests on science, in particular the science of evolutionary biology. Man was part of nature, and his societal instincts could be studied as one studies frogs or fish. "There is no cause," Peter claimed, "for suddenly changing our method of investigation when we pass from the flower to man or from a settlement of beavers to a human town."[22]

In the writings of Darwin, Kropotkin found the cohesive elements that linked mutual aid and politics in humans. In *The Descent of Man*, published twelve years after *The Origin,* Darwin offered explanations for why some human communities flourish and others fail. The answer was mutual aid. Kropotkin's reading of Darwin was that those communities which included the greatest number of altruists would flourish. Such was the real Darwin, as Peter understood him: "the chapter devoted by Darwin to this subject could have formed the basis of an entirely different and most wholesome view of nature and of the development of human societies."[23]

Kropotkin's efforts to promote a political system built on mutual aid were more than just a means of linking his political and scientific passions; he felt they were a matter of species survival. "We all know," Peter told the readers of *The Conquest of Bread,* "that without uprightness, without self-respect, without sympathy and mutual aid, human kind must perish, as perish the few races of animals living by rapine...."[24] Human survival was at stake, and its future depended on a new form of society. But how was it to be created? Kropotkin had both conceptual and practical answers.

Conceptually the challenge was straightforward. To achieve a good society, we must simply follow the rules of nature: "the course traced by the modern philosophy of evolution," as Kropotkin liked to call it. Once people understood the process of evolution, they could be convinced of the organismal aspects of society: an organism which was striving "to find out the best ways of combining the wants of the individual with those of co-operation for the welfare of the species."[25]

If such a society was studied the way that a natural historian would study, say, a land snail, the observations would necessarily also consider the danger of parasites. In human society, parasites came in the form of individuals who tried to suck the society dry, and co-opt all resources for themselves. Though these parasites threatened the very existence of the societal superorganism, they also provided a target. Eliminate the parasites, as well as the environment in which parasites thrived, and the result will be a society in which mutual aid reigns supreme.

The practical questions were how to achieve this, and where to begin. Kropotkin argued that anarchists should be focusing their attention on Western societies, where there was both an abundance of riches—leading to the possibility of a prosperous society—and a scourge of capitalist parasites. The riches of Western societies, Peter claimed, were accumulated through the hard work of millions of workers over long stretches of time. Early on, this involved building infrastructures, wherein "Every acre has its story of enforced labour," Peter lamented, "of intolerable toil, of the people's sufferings. Every mile of railway...has received its share of human blood."[26] As times changed, so too did the source of plenty. "With the *co-operation* of those intelligent beings, modern machines," Kropotkin wrote, "themselves the fruit of three or four generations of inventors, mostly unknown, a hundred men manufacture now the stuff to clothe ten thousand persons for a period of two years."[27] But the people whose work generated the wealth—those who actually built the infrastructure and designed the machines—shared little in the rewards that emerged from their work.

Economic parasites had stolen all the wealth—bought up the land and means of production, leaving the workers in squalor. The products of the masses had been usurped by a few cheaters, Peter proclaimed. "By what right," Peter asked, "can any one whatever appropriate the least morsel of this immense whole and say—this is mine, not yours?"[28] The unfair appropriation was in Peter's view unnatural; it ran counter to the evolutionary forces that drove societies toward acts for the good of all.

The solution to the human economic parasite problem was expropriation—the reallocation of riches from the few to the many. Only this would restore the evolutionary balance of nature. "This cannot be brought about by Acts of Parliament," Kropotkin wrote, "but only by taking immediate and effective possession of all that is necessary to ensure the well-being of all; this is the only really scientific way of going to work."[29]

Peter did not see violence as a necessary means to an altruistic end. The capitalist system was unnatural, and so it was, by definition, unstable, and easily rendered extinct. A combination of worker strikes and shutdowns, Kropotkin believed, would cause "the complete disorganization of the system based upon private enterprise and wage labour. Society itself will be forced to take production in hand, in its entirety, and to reorganize it to meet the needs of the whole people."[30] At that point, the people could take control of the distribution of food, clothing, housing and the means of production. It would take time, but freed from the shackles of the economic parasites, with the power of communal ownership, mutual aid would thrive. The superorganism that would result from this process "will not be crystallized into certain unchangeable forms," Kropotkin noted, "but will continually modify its aspect because it will be a living evolving organism."[31]

Such a society—in which "working 4 or 5 hours a day till the age of forty-five or fifty, man could easily produce all that is necessary to guarantee comfort to society"[32]—would educate all members and allow choice of careers. And the obvious benefits of mutual aid would encourage the people to pursue what was good for the community, setting up a mutual aid positive feedback loop.

Never one to see the limits of mutual aid, Kropotkin believed it not only drove successful societies, but provided a naturalistic explanation for the roots of human ethical behavior. Kropotkin found religion unsatisfying on this matter, and did not think that the Christian bible, or any religious text, would ultimately provide him an understanding of ethical behavior. He was fond of retelling a famous story in which Pierre Laplace presented Napoleon with a copy of his masterpiece, *The System of the World*. The Emperor was pleased, but noted that he had learned that there was no mention of God in the book, to which Laplace was said to reply "I had no need of that hypothesis."[33] Neither did Kropotkin.

The roots of ethics were unearthed by evolutionary biology. But not the dog-eat-dog evolutionary biology that "phony Darwinists" like Huxley, were promoting—ideas, Kropotkin argued, that led people to believe that "evil was the only lesson which man could get from Nature."[34] Instead, it was Darwin himself, who wrote of the evolution of human altruism and its effect on community in *The Descent of Species,* who provided Peter with a naturalistic explanation of ethics. "Being thus necessary for the preservation, the welfare, and the progressive development of every species," Kropotkin wrote in his magnus opus on the subject, entitled *Ethics: Origin and Development*, "the mutual-aid instinct has become what Darwin described as 'a permanent instinct,' ... is always at work in all social animals, and especially in man."[35]

Mutual aid, like all traits favored by natural selection, could be mapped onto an evolutionary tree. "Let us imagine," Kropotkin asked his reader, "how this feeling little by little became a habit, and was transmitted by heredity from the simplest microscopic organism to its descendants – insects, birds, reptiles, mammals, man." As soon as we can understand that, Peter proclaimed, we will understand ethical behavior, and once we do, we will see that it "is a necessity to the animal like food or the organ for digesting it."[36]

Over evolutionary time, organisms that acted ethically would be favored by natural selection. Kropotkin believed that animals and humans acted to maximize pleasure and minimize pain, and because mutual aid was beneficial to the species, natural selection would favor organisms that associated ethical action with pleasure, and the absence of ethical action with pain. "When a troop of monkeys has seen one of its members fall in consequence of a hunter's shot," wrote Kropotkin, "… these monkeys obey a feeling of compassion stronger than all considerations of personal security… This feeling becomes so oppressive that the poor brutes do everything to get rid of it."[37] And such ethical action was not limited to primates. Kropotkin's very next sentence contained the same impassioned rhetoric when it came to ants rushing into the flames to save their tiny comrades. "The animal world…from insects to man," Peter told his readers, "knows perfectly well what is good and what is bad, without consulting Bible or Philosophy."[38]

With Peter's science and politics touching on so many fundamental issues about humanity's place in nature, it wasn't surprising that people wanted to hear more of what he had to say. And so, it was time for Kropotkin to take mutual aid on the road. Happy to talk with anyone who would listen (and think), in August 1897, he set sail for the first of his two American speaking tours.

≡ CHAPTER 7 ≡

A WELL-PRESERVED RUSSIAN MALE

"It struck me personally as if he {Kropotkin} were speaking to us there in our poverty stricken Bronx flat."

Elizabeth Gurley Flynn[1]

Kropotkin the polymath toured North America in 1897 and again in 1901. There, over the course of many months and thousand of miles on the rails, often in front of packed houses whose audiences numbered in the thousands, Peter gave dozens of speeches on every subject from mutual aid and geology to Russian literature and Christianity. Other anarchists—Russian and otherwise—people, such as Michael Bakunin, Sergei Kravchinski,[2] and Alexander Berkman, had already visited or emigrated to North American by the turn of the 20th century, but the impact of their arrival on the Canadian and American people was mute compared to the reception Kropotkin received.[3]

Even before his first trip, Kropotkin was a well-known figure in both Canada and the United States. The story of his incredible escape from the Peter and Paul prison years earlier had been covered in newspapers, and his books, though often too technical for some readers, had been the rage both in intellectual circles, and amongst anarchists in late night coffee shop gatherings. His pamphlets, which were often more passionate and engaging, circulated as well. Peter's "An Appeal to the Young," was especially well-received. Elizabeth Gurley Flynn, who would one day become a leading figure in the American Communist Party and a cofounder of the American Civil Liberties Union, described the impact of this work on her own youth: "It struck me personally," Flynn wrote, "as if he were speaking to us there in our poverty stricken Bronx flat."[4] Emma Goldman, Peter's friend who emigrated to America from Russia in 1885, and was the most famous female anarchist of the 20th century, wrote that in 1890s America, Kropotkin was "recognized as one of the foremost men in the world…we saw in him the father of modern anarchism, its revolutionary spokesman and brilliant exponent of its relation to science, philosophy and progressive thought."[5]

Peter was first invited to visit the United States in 1891 by the Pioneers of Liberty, a group of German and Jewish anarchists located in New York. Kropotkin's reputation among Jewish anarchists in London was legendary, and his East End lectures drew terrific crowds. This attention no doubt piqued the interests of American Jewish anarchists like the Pioneers of Liberty. At the time, though, Peter was unable to accept the invitation, in part because he was feeling under the weather, and not in shape for a trans-Atlantic voyage.

It was science, not politics, that finally convinced Kropotkin to make the journey to North America. The British Association for the Advancement of Science (BAAS) was holding its annual meeting in Toronto in 1897, and the organizing committee invited Peter to speak on his work in geology. The chance to visit and talk about his research, along with the opportunity to spend some time with his friend, James Mavor, a radical socialist and Professor of Political Economics at the University of Toronto, was irresistible.

On August 24, 1897, Kropotkin spoke to the BAAS on what he saw as his crowning scientific achievement—the discovery of the correct orientation of the mountain ranges of Asia. Peter also took the opportunity to present a second talk, this one on glaciation patterns in Sweden and Northern Finland. By all accounts, these talks, each of which contained a mixture of hard science and a sprinkling of Peter's adventures in natural history, were well received. A small part of the attention they drew was no doubt due to the fact that in the meeting program, Peter was listed as *Prince* Kropotkin.[6]

Following his talk at the BAAS, Kropotkin boarded a Canadian Pacific Railroad train and embarked on a cross-country railroad excursion. Though Peter enjoyed scientific meetings, they could never compete with the thrill of such a cross-country journey. Joined by the director of the Canadian Geological Survey,[7] who knew the Canadian Rocky Mountains as well "as his own garden," Peter enjoyed every moment of the trip from Toronto to Victoria, British Columbia and then back again. The land itself was beautiful, Peter noted, and all along the way, people treated him with "the utmost cordiality and hospitality."[8] But it was more than that. Kropotkin was immersed in his favorite pursuits—geography and evolutionary biology.

Kropotkin found in the geography of Canada and the Northern United States a virtual a mirror image of that of his native Russia. "The traveller who would land in Russia on the coast of the Baltic" and then travel east, Peter noted, "would meet with exactly the same type of geographic regions…as those of which he meets crossing North America…but in the opposite direction."[9] To Kropotkin, this pleasing similarity was hardly a "fortuitous coincidence," but rather, evidence of the universal laws that governed geography and geology. The specifics of those laws were uncertain, but the overarching point for Peter was that they were universal.

Kropotkin wrote about the places he encountered on his trans-Canadian adventure with a sense of wonder and awe. "What a variety of landscapes," he told the readers of the magazine *Nineteenth Century*, "...the woody regions of the St. Lawrence...the cultivated hills and plains of Ontario...the boundless prairies...each of them a world in itself."[10] Peter fell in love with the Canadian West as an unexplored wonderland. "If time had permitted me to do so," he wrote, he would have visited "every settlement and town in the Northwest {Territories} and Manitoba."[11]

As much as the lands themselves, he admired the people that inhabited them. The Western Canadian communities and cooperatives exemplified mutual aid-based living. Peter encountered socialist-like cooperatives that produced vast quantities of dairy products, and he was constantly amazed at the ability of small communities to support themselves, one way or another, in what was often a dry barren landscape.

Kropotkin was especially taken by the Canadian Mennonites, "who prosper everywhere,"[12] and who often (but not always) operated as a single superorganism. In these Mennonite communities, there were common areas where cattle grazed under the eye of a communal shepherd, land was allocated to each family "in proportion to its working capacities," and a communal fund was established for buying more land for the community, when it was needed.[13] It was the Kropotkinian ideal, that had only one flaw—its reliance on religion, which Peter could never embrace. Still, any system of mutual aid, even one anchored on religiosity, was preferable to capitalism.

Revitalized from his adventures, and with what he called "a new lease on life,"[14] Kropotkin returned to Toronto from the Canadian West and rejoined his friend Professor Mavor. Conversations turned from his journey to socialism and communism. Peter described Mavor as "a virtual encyclopedia of Canadian economics,"[15] and he reveled in his company. While staying with Mavor, Kropotkin learned of an incident in Hazelton, Pennsylvania in which police had broken up a coal miners' strike,[16] wounding forty and killing twenty-one strikers. It was one of the few sad moments Peter experienced during his entire stay in North America—and he worried deeply that it was an omen of more violence toward union workers in the United States.[17]

Peter left Toronto shortly thereafter, in mid-October 1897, for a tour of the United States which began with a visit to anarchist Johann Most in Buffalo, New York. Johann thought of Kropotkin as not just "the celebrated philosopher of modern anarchism," but equally importantly to those who wanted anarchy to have a scientific footing, as "one of the greatest scientists of this century."[18] After spending a few days with Most, Peter traveled from Buffalo to Detroit to attend the annual meeting of the American Association for the Advancement of Science. His impressions from that meeting led him to predict that the United States would soon surpass Europe as the world's center for scientific learning. As he did at nearly every stop on his North American tour, Peter mixed politics and science while he was in Detroit. He spent time with Joseph Labadie, a socialist and anarchist, who had organized the Detroit Council of Trades and Labor Unions and served as its first president, and he took every opportunity to tour city projects, including the Detroit sewer system.[19]

After a quick stopover in Washington, where he gave a lecture about his Siberian expedition to the National Geographic Society, Peter moved on to Jersey City, New Jersey, where he was met by friends and an excited Press Corps, anxious to meet the Prince. When asked for a statement, Peter was pithy: "I am an anarchist," he told the newspaper men, "trying to work out the ideal society." The reporters were enamored with the Prince, describing him as "anything but a typical anarchist...he is patriarchal...his dress is careless, it is the carelessness of a man engrossed in science rather than that of the man who is in revolt against the usages of society."[20]

Across the Hudson river, in New York City, Kropotkin stayed with his anarchist friend John Edelmann on 96th Street and Madison Avenue. He spoke to an audience of trade unionists on "The Development of Trade Unionism."[21] Although his book *Mutual Aid* was still a few years off into the future, Kropotkin used this lecture to talk about medieval trade guilds that figured prominently in this book. The highlight of Kropotkin's visit to New York City was his October 24th lecture on "Socialism and Modern Development" at the city's famous Chickering Hall. All 2000 seats were sold, and despite bad weather the day of the lecture, an enormous crowd turned out for the Prince. Women present wore scarlet neckties to mimic Russian dress, and the Press was enamored. They described Peter as a "well-preserved gentleman of fifty-five years old,"[22] and were especially taken with the Prince's folksy approach, paraphrasing Kropotkin's claim "that if scientists were to clean sewers two hours a day they would invent so many improvements that it would become a comparative pleasure to work in sewers."[23] From Peter's perspective, the lecture went well, though tickets to Chickering Hall had been priced too high for the working man; from then on, he would insist that tickets to his talks be as cheap as possible.

Peter headed from New York City to Philadelphia on October 25, 1897. It was a brief stopover, in which he lectured about sociology and history to a crowd of 2000 at the Odd Fellows Temple. From Philadelphia, Kropotkin went to Boston, where he presented a series of talks at the Lowell Institute. These lectures—"Mutual Aid in Savages and Barbarians" and "Mutual Aid in The Medieval Society"—were particularly important to Peter, as they were the first public presentations of his ideas on evolution and mutual aid in North America.

Opening its door to progressive, imaginative thinkers since 1839, the Lowell Institute had hosted talks by such speakers as Charles Lyell (the founder of modern geology), Asa Gray (the founder of modern botany), Oliver Wendell Holmes, Alfred Russell Wallace (the co-founder of the theory of evolution, who gave eight lectures on "Darwinism and some of its applications") and psychologist William James (who lectured on "Exceptional Mental States").[24] The Lowell Institute was one of America's first true think tanks—exactly the sort of place in which Peter felt at home, and just the sort of venue he wanted to see spring up all across the world.

While Kropotkin was in the Boston area, his host was Charles Eliot Norton, a professor of the history of art at Harvard, and the first president of the Archaeological Institute of America. Norton had arranged for Peter to give a few talks besides the lectures at the Lowell Institute. These included a lecture on "The Socialist Movement in Europe" at Harvard, a talk to the Women's Industrial Club of Cambridge, and a pair of talks at a local church on "Christianity" and "Morality." By the time he left Boston for a return stop in New York City, Bostonians may have mused at a subject on which Peter Kropotkin could not lecture.

In November Kropotkin returned to New York City. There he met and talked politics with Saul Yanofsky, editor of the Yiddish Anarchist Newspaper, *Fraye Arbeter Shtime*. This stop in New York City also afforded Peter a rare pleasure—the chance to give a lecture in Russian on the "Philosophic and Scientific Bases of Anarchism." Kropotkin also availed himself of the opportunity to give a second talk at Chickering Hall, this time on "The Struggle for Freedom in Russia." His final lecture on the 1897 tour of North America was held at Cooper Union College, where Peter spoke on "The Great Social Problems of our Century." Kropotkin insisted that admission be set at a nickel, so that that the people who actually suffered the great social problems could attend.

Before Peter boarded the RMS *Majesty* to sail back to England, he was surprised again by his friend James Mavor. Mavor had used his contacts[25] to arrange for Kropotkin to write a series of autobiographical essays to be serialized in *The Atlantic Monthly*.

Eventually, these articles would become *Memoirs of a Revolutionist*, required reading for anarchists to this day.

Peter had much to do upon his return to England. There were anarchist pamphlets he needed to write on a suite of issues, and time also had to be set aside for the publication of his book *Fields, Factories and Workshops* (subtitled: *Industry Combined with Agriculture and Brain Work with Manual Work*), and his autobiographical essays for *The Atlantic Monthly*. But Americans had not had their fill of Kropotkin, and in 1901, he was invited to return for a second speaking tour by the Lowell Institute. As a testament to his intellectual prowess, and as evidence of his popularity, this time the Lowell Institute invited Peter to present a series of talks on Russian literature.

Kropotkin sailed to Boston in March 1901, and lectured at the Lowell Institute—these lectures were eventually compiled into yet another Kropotkin book, this one entitled *Ideals and Realities in Russian Literature*. During his month-long stay in the Boston area, Kropotkin spoke again at Harvard, and also gave a talk on "Anarchism: Its Philosophy and Ideal" before a very large audience at Paine Hall. Always leery of religious institutions, Peter initially declined an invitation from Edward Everett Hale, a well-known author, abolitionist, and clergyman, to speak at his church, but, eventually, Hale and Kropotkin agreed on a compromise—Peter would lecture at the church, but his talk would not be held in the sanctuary.

From Boston, Kropotkin embarked on his third visit to New York City on March 29, 1901. He was swamped by reporters at the Hotel Gerrard on his arrival there. He told them of the Russian secret police's crackdown on Russian universities, imploring the newsmen to get their readers to take some sort of action to help the embattled students.[26] Once he settled in at the hotel, Peter began a grueling series of lectures. He spoke about political economics to the Education League of New York City and about Tolstoy's writings to a packed house at The Berkeley Lyceum.[27] He lectured on "Anarchism: Its Philosophy and Ideals" to 4000 people at The Grand Central Palace, one of the largest speaking auditoriums in New York City.[28] He even managed to find time to give a lecture in Russian at New York's Tammany Hall.

In perhaps the strangest encounter of his life, while he was in New York City, Kropotkin was taken by his friend, Robert Ely, to meet with 75-year-old Varina Davis, the widow of Confederate President Jefferson Davis. Ms. Davis had requested the private interview, to which Peter was happy to oblige. While they spoke in a hotel room, Booker T. Washington appeared in the lobby of the hotel, in search of Robert Ely. When Ely learned of this, he arranged for Washington to join Davis and Kropotkin. No record of their conversation exists, but it must have been quite the discussion. Peter's daughter subsequently claimed that it was during this meeting that Kropotkin convinced Washington to write his autobiography.[29]

Exhausted, Peter came down with the flu and was bedridden for a week. When he recovered in early April, Kropotkin began his first tour of middle America. His initial stop was scheduled to be Chicago, but on the train ride out, he decided to stop in Pittsburgh to visit his imprisoned friend Alexander Berkman. Berkman, a leading anarchist of the day, was serving a term for the murder of Henry Clay Frick, who had led a violent suppression of strikers in the famous Homestead Steel strike of 1892. Kropotkin was denied visitation with Berkman, who several days earlier had been thrown into solitary confinement for attempting to tunnel his way out of his cell. Even without a meeting, Peter managed to infuriate the prison warden by writing a letter a few days later and addressing it to "Alexander Berkman, Political Prisoner."[30]

When he arrived in Chicago, Peter stayed at Jane Addam's "Hull House," the social hub of the city.[31] Hull House had already served host to many of the more controversial figures to visit Chicago, but "In this wonderful procession of revolutionists," Addams wrote, "Prince Kropotkin, or as he prefers to be called Peter Kropotkin, was doubtless the most distinguished."[32] In honor of her famous guest, Addams had Hull House decorated in a Russian motif, and had her staff don Russian costumes for his entire stay. Though some anarchists were not pleased that Kropotkin was hosted by "the porkocracy of Chicago,"[33] Peter enjoyed his stay at Hull House, and presented a lecture there on "Fields, Factories and Workshops."

As he settled into Chicago, Peter presented a series of talks on "The Law of Mutual Aid and the Struggle For Existence" at the Twentieth Century Club. On April 20th, he took time from his lecturing to lay flowers on the graves of the anarchists that had been killed at the infamous Haymarket Riots in 1886. Then it was back to the lecture circuit, where Kropotkin spoke on "Science and the Social Question" at the University of Chicago, and "Anarchism: Its Philosophy and Ideal" before a crowd numbering three thousand at the Central Music Hall.

Peter's tour of the American Midwest next took him to the Champaign-Urbana area of Illinois. On the train ride from Chicago, Kropotkin spent his time gazing out his window, collecting data on American farming techniques. When he arrived at Champaign-Urbana on April 23, he spoke at the University of Illinois on "The Modern Development of Socialism," and then boarded a train and headed to Madison, Wisconsin, where he presented his last American lecture at the University of Wisconsin.

From Madison, Kropotkin headed east, traveling through Indiana and Ohio, taking notes on American farms and farming techniques. He made a stopover in Buffalo to spend a few days with James Mavor, whom he had not seen since his first trip to America in 1897. Then, in early May, 1901, Peter boarded a ship from New York City and headed back home to England. Soon after he arrived, he suffered a mild heart attack, which he attributed to his relentless American schedule.[34] But he was already planning his third visit to America.

Four months after Peter left America for the second time, President William McKinley was assassinated at the Pan-American Exposition in Niagara Falls, New York by anarchist Leon Czolgosz, who told investigators, "I killed President McKinley because I done my duty. I didn't believe one man should have so much service and another man should have none." A strong anti-anarchist tide quickly swept across America. Though not a stitch of evidence supported the claim, rumors began to circulate that Kropotkin, along with Emma Goldman, had hatched the McKinley assassination plot while Peter was at Hull House. Soon thereafter, as a direct response to the killing of the President, the Congress of the United States passed the Immigration Act of 1903, which extended the list of banned immigrants to include "anarchists, or persons who believe in, or advocate, the overthrow by force or violence the government of the United States, or of all government." There would be no third visit to America for Peter Kropotkin.

CHAPTER 8

THE OLD FOGEY

"From the point of view of liberty, what system would be best? In what direction should we wish the forces of progress to move? ... I have no doubt that the best system would be one not far removed from that advocated by Kropotkin."

Bertrand Russell[1]

Back in London in 1901, Peter struggled to earn a living. He was having difficulty finding a publisher for his new book, *Ideals and Realities in Russian Literature*, and was still recovering from his heart attack, which limited his ability to work. He warned friends, "I know I cannot live long... ."[2] Peter's lectures in North America had been an unabashed success, but a financial nightmare. Most of his hosts could cover train fare, and room and board, but provided little or no honorarium. On the rare occasions when Peter was well paid for an event, he would usually donate his honorarium to a cause, such as resettling the oppressed Dukhobors people of Russia to new settlements in Canada.

Kropotkin was also frustrated at his stalled mutual-aid-based anarchist revolution. The cause wasn't helped by the rising number of assassinations of political leaders, some claimed at the hands of so-called anarchists. Though he refused to publicly condemn specific acts of terrorism if the terrorists were part of an oppressed group, Kropotkin viewed single acts of terrorist violence as both barbaric and counterproductive to the anarchist cause. Rather than support the terrorist faction of various anarchist groups around the world, Peter aligned himself with the syndicalists—a union group similar to the medieval guilds that he so adored.

Anarchist societies weren't thriving, and a rival political theory was gaining ground at the expense of Peter's biological anarchism. Though both anarchism and Marxism shared the general goal of an equitable distribution of resources, Peter abhorred the approach Marxism took to achieve this goal. Kropotkin considered himself a true communist—in the sense of the commune as a group of individuals sharing common property and income, with nonhierarchical decision-making bodies—not a Marxist, with a State-centered solution to all problems big and small. Small, cooperative, autonomous, but connected societies freed from government, where "each region will become its own producer and its own consumer of manufactured goods," were the only path toward a better world.[3] Peter's life experiences, and his knowledge of evolutionary biology and philosophy, had convinced him that in the State lay the problem, not the solution.

Marx, like Kropotkin, had tried to wrap his political theory in biology, but Peter would have no part of that. Kropotkin argued that Marxism was biologically shallow, as it focused solely on human societies and their dynamics: it was too narrow to be useful. Instead, it was anarchy via mutual aid that represented the only true link between politics and biology. "Its {Anarchism} method of investigation is that of the exact sciences," Peter wrote, "...its aim is to construct a synthetic philosophy comprehending in one generalization all phenomena of Nature."[4] Marxism, with its State-centered philosophy and its narrow biological focus, could not achieve this. What's more, Peter also saw Marxism as a cult. "Their Mecca is Berlin," Kropotkin wrote, "their catholic religion is Marxism. As for the rest—I don't give a damn."[5] And, yet, Marxism, was clearly a serious threat to Peter and the anarchists at the turn of the 20th century, and contributed to Peter's woe.

Then, in 1905, came the Russian peasant revolt, the result of the Czarist government's clamping down on workers who had been striking more and more often in Saint Petersberg and Central Russia, including a massive strike of 120,000-150,000 in St. Petersberg in January 1905. "Bloody Sunday" marked the start of the revolt.[6] Though Peter abhorred the loss of life at the hands of armed troops, the 1905 revolution reinvigorated him. This was a leaderless, nonviolent revolution of the people, small groups working together to create the ideal society. Precisely the revolution that Peter the scientist and Peter the political philosopher had been working for all these years. At last, the mutual aid he had seen amongst the animals in the Siberian tundra had filtered its way up the species chain to the political system.

Kropotkin became obsessed with following what was happening after Bloody Sunday, and was delighted when the strikes spread across Russia. He wrote and spoke about the peasant revolution, including an article in *Khleb I Volia*, the first anarchist newspaper to be published in Russian. He began to attend anarchist conferences, not just in London, but in Paris, where he was still technically classified as an agitator, but was left alone by the French authorities.[7]

Enthralled with the emergence of mutual aid and anarchy in his very homeland, Peter contemplated a return to Russia. He went so far as to pick up a rifle and engage in some target practice to see if he could be useful should violence become necessary. Ultimately, advice from friends, combined with his poor health, and the almost certain arrest that would follow his return, convinced Peter to stay in England, much to his dismay: "If only I were somewhat younger," he wrote a friend, "I could live underground."[8]

Though the Czar made some concessions in response to the 1905 revolution,[9] within two years the Russian government had largely regained its absolute grip on power.[10] Kropotkin's resolve strengthened in response to the failed revolution, vowing to insure the success of future uprisings that would lead to an anarchist Russia steeped in the principles of mutual aid. His efforts took the forms of speeches and papers on mutual aid, anarchism, and the political and humanitarian conditions in Russia.

By the first decade of the 20th century, Kropotkin was aware of his celebrity, and decided to use his renown on the world stage to keep the international spotlight focused on Russia and the brutal Czarist regime. Fortunately, his finances had improved, via a combination of increased book sales and downsizing to a smaller house in Brighton, leaving him the time and wherewithal to accomplish his mission.

In a seventy-five page pamphlet entitled *The Terror in Russia*,[11] Kropotkin starkly laid out the Czar's response to attempts at revolution, pleading for help from the citizens of the world. "The present conditions in Russia are so desperate," Peter began, "that it is a public duty to lay before this country a statement of these conditions, with a solemn appeal to all lovers of liberty and progress for moral support in the struggle that is now going on for the conquest of political freedom… we should not forget that there exists a web of international solidarity between all civilised countries." He intended to elicit such international help by shocking the reader of *The Terror in Russia* with data on such atrocities as the widespread execution of innocent peasants.

Kropotkin devoted pages of this pamphlet to the fetid conditions of the jails housing political prisoners, including a chilling list of the names of 160 prisoners who had committed suicide, one of which was a Ms. T. Savitzkaja, who, at the orders of the Czar, was housed in "the common cell for women prisoners" in Odessa, and who "cut her throat and stomach with a piece of glass."[12] Kropotkin pleaded with his readers to keep international pressure on Russia to stop such political repression: "To all those who realise the unity of mankind," Peter ended his pamphlet, "this exposure of the horrors of the present repression in Russia is sure to appeal."[13]

Peter understood that even if the Czarist government were deposed, such a course of events wouldn't necessarily lead to a mutual aid-based society. To see that it did, he continued to write on the philosophy and morality of anarchism, including a very high profile entry on "Anarchism" in the 1910-1911 edition of the *Encyclopedia Brittanica*. Anarchy, Peter wrote, was the "principle or theory of life and conduct under which society is conceived without government." In anarchist societies, which were a kind of "organic life at large," Peter told his reader, "harmony {is} obtained, not by submission to law, or by obedience to any authority, but by free agreements concluded between the various groups...for the satisfaction of the infinite variety of needs and aspirations of a civilized being," including, of course, "the satisfaction of an ever-increasing number of scientific, artistic, literary and sociable needs."[14]

Though the *Encyclopedia Brittanica* entry was not about Russia per se, Peter's message to the reader regarding Russia and revolution was clear: "The anarchists recognize," he wrote, "that like all evolution in nature, the slow evolution of society is followed from time to time by periods of accelerated evolution which are called revolutions; and they think that the era of revolutions is not yet closed."[15] More such revolutions, Peter was certain, were to follow in his homeland.

With his philosophy of mutual aid as his guiding principle, Kropotkin viewed nonviolent revolution as the path to an anarchist society. Certainly people would die in such a revolution, but large-scale violence was not part of Kropotkin's revolutionary strategy. He was not a pacifist per se, but, as one might expect from the world's leading spokesman for the biological theory of mutual aid,

Peter believed large-scale violence rarely achieved any moral good. His convictions, though, were about to be challenged by the events of World War I.

For Kropotkin, all State-run governments were impediments to the spread of mutual aid, but some were much greater impediments than others. German Nationalism of the early 1900s was, in his mind, the greatest threat. Radical German nationalism, coupled with virulent militarism, not only threatened the stability of world politics (not always a bad thing in Peter's eyes), but was antithetical to any system of mutual aid. In one of the few instances in which Peter adopted an "ends justify the means" approach, he argued that Germany had to be stopped, with massive military force if necessary.

During World War I, Kropotkin strongly supported the Allies in their fight against Germany and other Axis countries, going as far as signing a very public document called the Manifesto of 16, in which he and others called for the defeat of Germany at all costs. Almost all other leading anarchists adopted pacifist positions, arguing that war was just a means for the State to usurp more power. Kropotkin appreciated this stand, having adopted it himself many times in the past, but German nationalism was so great a threat to mutual aid, he felt compelled to support the war. This caused a rift between Peter and many of his oldest and dearest friends—a rift which never healed.

Germany was threatening Peter's homeland. His support of the war had cost him many dear friends, he was approaching 40 years in exile from Russia, and he was still ill. His doctor had ordered him bed-bound for months, after which he could only be pushed in a wheelchair around his Brighton home for months more. As he recovered, though, Peter learned of the February 1917 revolution in Russia, which deposed the Czar and the Romanov dynasty once and for all. It was almost too good for Peter to believe.

The February revolution was leaderless, and had been relatively peaceful, the result in part of strikes, protests and student-led resurrections. Even more remarkably, the army joined the revolution, insuring the demise of Czarist rule. The Provisional Government that took the reigns of leadership[16] was still a government, but infinitely preferable to the autocratic and brutal regime of Czar Nicholas II. The image of a new flag flying over the dreaded Peter and Paul Prison that housed him so many years earlier left Peter delighted, recounting to an Associated Press reporter, "{I} regard the fall of autocracy as final."[17]

Peter could no longer miss the first-hand experience of a new Russia. He and Sophia packed all their belongings, including 50 crates of books, and traveling under the assumed name of Mr. and Mrs. Sergei Tiurin, sailed back home. They stopped briefly in Sweden and Finland, where they were greeted by hundreds of soldiers, including a military band playing "La Marseillaise."

The Kropotkins arrived in Petrograd on May 30, 1917, and were met by a crowd of thousands that were expecting them,[18] including two government ministers (Kerensky and Skobolev). Soon after their arrival, Prime Minister Kerensky offered Peter the position of Minister of Education in the Provisional Government,[19] but Kropotkin declined, citing his distaste for State government. Though he refused any official position, Peter did work with Kerensky to try and secure resources for those in need in Russia.

By August, the Kropotkins had moved from Petrograd to Moscow. Two months later, Kropotkin's dreams of an anarchist Russia were quashed for the final time. With the October 1917 revolution, and the rise of Lenin and his tightly-run, completely centralized State government, Russia was now headed in precisely the opposite direction that Kropotkin had worked so hard to secure.

After the October revolution, Kropotkin and Lenin exchanged a number of letters and even met on at least one occasion. Peter appealed to Lenin to allow the people to govern themselves, to secure resources for Russia's starving peasants, and to cease the execution of enemies of the State. Lenin was cordial in his replies, even asking Peter if he would like his book on the French Revolution published by a State press, but he honored none of Kropotkin's requests. Indeed, Lenin viewed his exchanges with Peter as a way to garner some support from the peasants, and little more. In private, Lenin was not as cordial, writing "I am sick of this old fogey {Kropotkin}. He doesn't understand a thing about politics and intrudes with his advice, most of which is very stupid."[20] Trotsky, too, had little use for Kropotkin, dismissing him as a "superannuated anarchist."[21]

In June 1918, The Kropotkins moved to a tiny house in Dmitrov, a village 60 kilometers north of Moscow. Rumors were afloat that they had been imprisoned there by the Bolsheviks, but there was no truth to such claims: indeed, Peter had actually been offered a professorship in geography at Moscow University, but declined because he had no interest in moving back to Moscow.[22]

Food was scarce, but better than for most, as the Kropotkins received extra "academic rations." Though Dmitrov was hard to reach from Moscow, visitors to the Kropotkin abode were plenty. Friends such as Emma Goldman and Alexander Berkman came by to check on their great leader, and foreigners, such as *New York Times* reporter Herman Bernstein, and Edgar Sisson, a representative sent by President Wilson, would make their way to Dmitrov to meet Peter and get a statement from him about his thoughts on Russia's future.

Though Kropotkin knew he would never live to see the anarchist world of which he dreamed, he continued with his work on mutual aid and anarchy. Health problems prevented him from accepting invitations to lecture on these subjects outside of Dmitrov, but he was able to secure a used typewriter, and some hard-to-come by paper, and spent much of his time in his study, writing *Ethics: Origin and Development*, his magnum opus on the role of mutual aid in structuring animal and human societies—what Peter called, his "great book on ethics on a naturalistic basis."[23] Emma Goldman, who visited Kropotkin in 1920, described him writing "his *Ethics* by the flicker of a tiny lamp that nearly blinded him. During the short hours of the day he would transcribe his notes on a typewriter, slowly and powerfully pounding out every letter."[24]

As always, writing about mutual aid didn't satisfy him: Peter needed action. He became an integral part of the Dmitrov cooperative, a small peasant-run group that modeled itself after the structures Kropotkin had written of so often. Peter's primary duty in the Dmitrov cooperative was to act as curator for a very small geology collection—rocks, minerals, and so on—that the people of Dmitrov had amassed. He also had the joy of lecturing to cooperative members on his adventures in Siberia almost fifty years earlier.

On October 15, 1920, the *New York Times* published a story entitled "Kropotkin is Starving." The paper had learned from a German trade union that "the veteran political fighter" was in "want of food" and that "his death is practically certain during the coming Winter."[25] In late January of that winter, the paper reported that Peter had died, only to retract their story and apologize for the error five days later.[26] A week after that, on February 9, 1921, the *New York Times* reported, correctly this time, that Peter Kropotkin was dead.

Peter's body was sent, via train, to Moscow for a funeral ceremony set for February 12th.[27] Lenin offered to have a State funeral, but Peter's family declined. Instead, a collection was quickly taken up by local anarchist groups to fund the ceremony. A huge crowd, numbering in the thousands, met the train carrying Peter's body. They slowly marched the coffin to the Palace of Labor, where it rested in the Hall of Columns.

At the Palace of Labor, the orchestra from the Moscow Opera played pieces from Kropotkin's favorite composer. Many of the groups that attended the funeral brought flags representing their organizations: banners of anarchist parties, trade unions, science societies, and literary societies filled the hall.

Kropotkin had, in fact, been in the Palace of Labor once before, when it was called the Palace of Nobility, and Peter, a terrified eight-year-old boy in a Persian Prince costume, was called up to meet the Czar, who proclaimed aloud that "this is the sort of boy you must bring me."[28]

≋ EPILOGUE ≋

In the late 1980s, while researching my own Ph.D. dissertation on animal behavior and the evolution of cooperation, I came across many citations to Peter Kropotkin's work on this same topic. Quickly I came to realize a few things. Either these citations were "throw aways"—that is, citations to books the authors themselves had never read—or they were about Kropotkin the anarchist, not Kropotkin the scientist. But, when I read Kropotkin's books, cover to cover—which I did many times, in part because they are so wonderfully written—I realized that his ideas were so much more important than indicated in the evolution and animal behavior literature that I was voraciously reading as a graduate student, and continue to consume to this day.

In addition to Kropotkin being one of the most famous political anarchists in history, he was an extraordinarily important figure in terms of his science. He was the first person to propose that animal cooperation was crucial for understanding the evolutionary process. He challenged the prevailing Darwinian principle that evolution was strictly about survival of the strongest. It would have been remarkable enough if Kropotkin had done this in obscurity, but quite the contrary—in his day he was *the* public face of these ideas, and one of the most recognizable people on the planet, lecturing on an astonishing array of subjects all over the world.

There is currently an entire subdiscipline in biology devoted to the study of cooperation and altruism in animals. This is not a small enterprise. E.O. Wilson called understanding animal cooperation and altruism one of the fundamental problems in the study of animal behavior, and that emphasis can be seen in the laboratories of scores of researchers who specialize in this area today—laboratories from UCLA to Princeton, from the University of Texas to the University of Helsinki. Kropotkin's work in the late 1800s marks the birthplace of this field.

Many of the ideas that are the focus of research in modern labs working on animal cooperation are based on permutations of ideas first raised to the surface by Peter Kropotkin. Literally hundreds of papers come out each year on animal cooperation—many in preeminent journals such as *Nature* and *Science*—and so many of these papers show Kropotkin to be a prophet. Kropotkin spoke of cooperative animals on sentinel duty: today labs at Cornell and Cambridge have dozens of people in the field studying this, and they have Kropotkin to thank for bringing the subject to the floor, and convincing people just how important it was.

And Kropotkin was not only the first person who clearly demonstrated that cooperation was important among animals, he was the first person to forcefully argue that understanding cooperation in animals would shed light on human cooperation, and, indeed would permit science to help promote human cooperation, perhaps saving our species from destroying itself. Today, anthropologists, political scientists, economists and psychologists publish hundreds of studies each year on human cooperation, and researchers in these fields are just beginning to realize that so many of the topics they are investigating were first suggested and promulgated by Peter Kropotkin.

Comrade farewell! Your life was not indeed
A martyrdom consoled
Maternally by Death: You lived to bleed
Your years growing old
Captive or exile, steadfast and unfurled
For Russia and World!...

That virile, Mosaic beard, those glances keen
The famed clasp of your hand
Express a life shriven and as clean
As cleanly ocean sand
Tomorrow's dawn lights up your kindly beard
And who would call you dead?

Russia unbars her gates; nightly barely over
She does not see—she feels
Her groping spirit yearns to know its lover
Sly footfalls dog her heels
Your grave is Russia's bread. Peace to the two!
Peace to Russia – and you!

A poem by Rose Florence Freeman to mark the occasion of Kropotkin's death[342]

ACKNOWLEDGEMENTS

I am deeply indebted to Christie Henry, whose advice and friendship have been priceless. I'd also like to thank all of the anonymous reviewers who provided insight, criticisms and suggestions on this book. The librarians at the International Institute of Social History in Amsterdam were extremely gracious when I visited, making available all documents in their extensive Kropotkin collection. My visit to the IISH was funded by a gracious grant from the University of Louisville.

As always, my wife Dana has read the entire manuscript more times than anyone should have to and her thoughts and suggestions have made the book that much better. She also makes my life much better. Lastly, thanks to my son Aaron, who despite his misguided sense that the New York Yankees are only the second best team in baseball, makes my days so much more enjoyable.

Chapter Notes

Notes to Preface

1. Wilde, O. *The Complete Works of Oscar Wilde*, Oxford University Press, 2005, p. 124.
2. Kropotkin, P. *Mutual Aid: A Factor of Evolution*. Third ed. London: William Heinemann, 1902, p. xxxvii.
3. Kropotkin, P. *Memoirs of a Revolutionist*. Boston: Houghton Mifflin Company, 1899, p. 215.
4. *Memoirs,* p. 216.
5. *Memoirs,* p. 217.
6. Kropotkin, P. Charles Darwin, Le Revolte, April 29, 1882.

Notes to Chapter 1

7. Rose Florence Freeman's poem can be found in: Ishill, J. *Peter Kropotkin, the Rebel, Thinker and Humanitarian*. Berkeley Heights, NJ: The Free Spirit Press, 1923.
8. Brooke, C. *Moscow: A Cultural History*. New York: Oxford University Press, 2006, p. 186.
9. Known as the Staraya Konyushennaya.
10. The street on which he was born is now named for Kropotkin, Kropotkinsky pereulok; Brooke, 2006.

11 For more on Peter's early life, see: Miller, M. A. 1967. The formative years of P.A. Kropotkin, 1842-1876: A Study of Origins and Development of Populist Attitudes in Russia. Ph.D. Thesis. Department of History, University of Chicago; Rogers, J. A. 1957. Prince Peter Kropotkin, Scientist and Anarchist: A Biographical Study of Science and Politics in Russian History. Ph.D. Thesis. Department of History, Harvard University; Miller, M.A. *Kropotkin*. Chicago: University of Chicago Press; Kropotkin, P. 1899. *Memoirs of a Revolutionist*. New York.
12 Miller, 1976, p. 3.
13 *Memoirs*, p. 8.
14 Though he was likely involved in the Turkish campaign of 1828.
15 *Memoirs*, p. 9.
16 A good example of Alexander's attitude toward the military is recounted by Peter as follows: "if we children, taking advantage of a moment when he was in a particularly good temper, asked him to tell us something about the war, he had nothing to tell but of a fierce attack of hundreds of Turkish dogs which one night assailed him and his faithful servant, Frol, as they were riding with dispatches through an abandoned Turkish village. They had to use swords to extricate themselves from the hungry beasts. Bands of Turks would assuredly have better satisfied our imagination, but we accepted {this} as a substitute. When, however, pressed by our questions, our father told us how he had won the cross of Saint Anne "for gallantry," and the golden sword which he wore, I must confess we felt really disappointed. His story was decidedly too prosaic. The officers of the general staff were lodged in a Turkish village, when it took fire. In a moment the houses were enveloped in flames, and in one of them a child had been left behind. Its mother uttered despairing cries. Thereupon, Frol, who always accompa-

nied his master, rushed into the flames and saved the child. The chief commander, who saw the act, at once gave father the cross for gallantry "But, father," we exclaimed, "it was Frol who saved the child. What of that? " Replied he, in the most naive way. 'Was he not my man? It is all the same.'" *Memoirs*, p. 10.
17 Miller, 1967, p. 12.
18 *Memoirs*, p. 12-13.
19 *Memoirs*, p. 28-31.
20 *Memoirs*, p. 24.
21 *Memoirs*, p. 25.
22 *Memoirs*, p. 25.
23 *Memoirs*, p. 18.
24 *Memoirs*, p. 44.
25 Miller, 1976, p. 16-18; Miller, 1967, p. 48.
26 Miller, p. 20.
27 The Corps had the reputation of being a brutal place: "a thousand blows with birch rods were sometimes administered, in the presence of all the corps, for a cigarette—the doctor standing by the tortured boy, and ordering the punishment to end only when he ascertained that the pulse was about to stop beating. The bleeding victim was carried away unconscious to the hospital." *Memoirs*, p. 54.
28 Peter to Alexander (Sasha) Kropotkin. Lebedev, M. N. 1932. Petri Aleksandr Kropotkiny: Perepiska. Academia, Moscow, volume I, p. 59.
29 Peter to Alexander (Sasha) Kropotkin. Lebedev, M. N. 1932. Petri Aleksandr Kropotkiny: Perepiska. Academia, Moscow, volume I, p. 5.
30 Levshin, D. 1902. Pazheskii korpus ego imperatorskoge velichestva. Za sto let, 1802-1902, St. Petersburg., p. 563, as cited in Miller, 1967, p. 58.
31 Miller, 1967, p. 69.

32 More on this correspondence can be found in Slatter, J. 1994. The correspondence of P.A. Kropotkin as historical source material. The Slavonics and East European Review 72:227-288. The letters are collected in: Lebedev, N. 1932.
33 *Memoirs*, p. 97.
34 *Memoirs*, p. 117.
35 *Memoirs*, p. 124.
36 Kropotkin wrote about the gloom that descended on Moscow during the siege at Sebastopol; *Memoirs*, pp. 63-64.
37 *Memoirs*, pp. 115.
38 *Memoirs*, pp. 115.
39 *Memoirs*, pp. 141.
40 *Memoirs*, pp. 147.
41 *Memoirs*, pp. 150.
42 Peter first mentions the Amur region in September 1858. Miller, 1967, p. 135.
43 *Memoirs*, pp. 155.
44 *Memoirs*, pp. 166-167.

Notes to Chapter 2

1. *Memoirs*, p. 239.
2. Miller, 1976, p. 53.
3. *Memoirs*, p. 168.
4. Woodcock, G., and I. Avakumovic. 1950. *The Anarchist Prince*. A Biographical Study of Peter Kropotkin. T.V. Boardman and Co., London, p. 56.
5. Kropotkin, P. 1883. Outcast Russia. Nineteenth Century December 1883:963-976, p. 964.
6. Miller, 1976, p. 59.
7. Lonergan, L. 1996. M.L. Mikhailov and Russian radical ideas about women, 1847-1865. Ph.D. thesis, University of Bristol; Woodcock, G., and I. Avakumovic. 1950, p. 57

8. Frank, J. 1988. *Dostoevsky: The Stir of Liberation*. Princeton University Press, p. 137
9. In Siberia, Kropotkin would also have contact with "Decembrists" D.I. Zavalishin and I.I. Gorbachevskii; Miller, p. 59.
10. *Memoirs,* p. 213.
11. *Memoirs,* p. 198.
12. *Memoirs,* p. 194-195.
13. *Memoirs,* p. 192.
14. *Memoirs,* p. 168.
15. *Memoirs,* p. 168.
16. *Mutual Aid: A Factor of Evolution,* xxxv.
17. *Mutual Aid: A Factor of Evolution,* xxxv.
18. *Mutual Aid: A Factor of Evolution,* xxxiv-xxxv.
19. *Mutual Aid: A Factor of Evolution,* xxxvi.
20. *Mutual Aid: A Factor of Evolution,* xxxvii.
21. *Mutual Aid: A Factor of Evolution*, xli. Although Kropotkin the scientist arrived in Siberia with a fresh impression of Darwin's *Origin of Species,* and expected to find much competition between animals, Peter the passionate philosopher may have been predisposed in the opposite direction. In his *Memoirs,* Peter wrote of a story he read of one of his intellectual idols, Goethe: "Eckermann told once to Goethe...that two little wren-fledglings, which had run away from him, were found by him next day in a nest of robin redbreasts, which fed the little ones, together with their own youngsters." Peter continued the tale: "Goethe grew quite excited about this fact and said: 'If it be true that this feeding of a stranger goes through all Nature as something having the character of a general law—then many an enigma would be solved.'" Even here, in his discussion of Goethe's anecdote, Kropotkin could not help but note the scientific suggestion with which it concluded: "He {Goethe} returned to this matter on the next day, and

most earnestly entreated Eckermann...to make a special study of the subject, adding that he would surely come to quite invaluable treasuries of results." Unfortunately, Peter informed his own readers, such a study had never been done.

22. *Memoirs,* p. 215.
23. *Memoirs,* p. 216.
24. His papers on "Mutual Aid in Savages", "Mutual Aid in Barbarians," and "Mutual aid in the Medieval City" will be discussed at length in a later chapter.
25. *Memoirs,* p. 216.
26. Miller, 1976, p. 37.
27. *Mutual Aid: A Factor of Evolution,* xli.
28. *Mutual Aid: A Factor of Evolution,* xxxv.
29. *Memoirs,* p. 199.
30. *Memoirs,* p. 200.
31. Kropotkin, P. 1904. The orography of Asia. Geographical Journal February-March 1904:4-61.
32. *Memoirs,* p. 226.
33. *Memoirs,* p. 226.
34. *Memoirs,* p. 226.
35. *Memoirs,* p. 227.
36. *Memoirs,* p. 226.
37. *Memoirs,* p. 216.
38. *Memoirs,* p. 168.
39. Kropotkin, P. 1887. The scientific bases of anarchy. Nineteenth Century 21:238-252; p. 238.
40. *Mutual Aid: A Factor of Evolution,* p. 186.
41. Kropotkin, 1887, The scientific bases of anarchy, p. 238-239.
42. *Memoirs,* p. 403.
43. Technically, Kropotkin was still in the service of the Czar, though in a nominal position in the Ministry of the Interior. He resigned from the Ministry of the Interior in 1872; Woodcock, p. 76.

44. *Memoirs*, p. 224.
45. Miller, 1976, p. 73; Kropotkin, P. 1896. The Commune of Paris. Liberty Library.
46. Miller, 1976, p. 72.
47. *Memoirs*, p. 234-236.
48. *Memoirs*, p. 237-8.

Notes to Chapter 3

1. *Memoirs*, p. 343.
2. *Memoirs*, p. 303.
3. *Memoirs*, p. 242.
4. *Memoirs*, p. 247-8.
5. Kropotkin was introduced to the Tchaykóvsky Circle by Dmitri Klemens.
6. McKinsey, P. 1974. The Chaikovskii Circle and the Origins of Russian Popularism. Ph.D., The University of Missouri, Columbia.
7. And to a much lesser extent the narodniks: members of the middle class.
8. *Memoirs*, p. 304.
9. According to one Tchaykóvsky Circle member named Perovskaia: Miller, 1976, p. 90.
10. Woodcook and Avakumovic, 1950 p. 124.
11. After his father's death.
12. *Memoirs*, p. 302.
13. *Memoirs*, p. 302.
14. *Memoirs*, p. 310.
15. As well as the narodniks.
16. *Memoirs*, p. 333.
17. *Vperdi*, 1874, III, p. 247, as cited in Miller, 1967, p. 323-324.
18. Until 1870 this was known as the Peter and Paul Fortress.
19. *Memoirs*, p. 343.
20. *Memoirs*, p. 341.

21. *Memoirs,* p. 345
22. *Memoirs,* p. 345.
23. *In Russian and French Prisons, p.* 91-93; Kropotkin, P. 1883. The fortress prison of St. Petersburg. Nineteenth Century XIII:928-943.
24. *Memoirs,* p. 346.
25. *Memoirs,* p. 345.
26. *Memoirs,* p. 345.
27. *Memoirs,* p. 351.
28. *Memoirs,* p. 350.
29. *Memoirs,* p. 350.
30. Kropotkin, P. 1904. The orography of Asia. Geographical Journal February-March 1904:4-61.
31. Grapes, R., and D. Oldroyd. 2008. History of geomorphology and quaternary geology. Published by the Geological Society.
32. Kropotkin envisioned two reports/books. The second manuscript "remained in the hands of the Third Section when I ran away. The manuscript was found only in 1895, and given to the Russian Geographical Society, by whom it was forwarded to me in London." *Memoirs, p. 351.*
33. *Memoirs,* p. 364.
34. Miller, 1976, p. 119, fn 21.
35. Miller, 1976, p. 119, fn 22.
36. Also known as the Nikolaievsk Military Hospital. Kennan, G. 1912. The Escape of Prince Kropotkin. The Century Magazine 62:246-253. *The New York Times* discussed Kennan's description of the Kropotkin escape: *The New York Times* March 7, 1915.
37. *Memoirs,* p. 365.
38. Kropotkin, P. 1887. *In Russian and French Prisons.* Ward and Downey, London, p. 99.
39. *Memoirs,* p. 367.
40. *Memoirs,* p. 368.
41. *Memoirs,* p. 367-8.

42. Old style calendar.
43. *Memoirs,* p. 369.
44. Kennan, 1912. George Kennan (1845 – 1924) was an American explorer known for his work in Russia. While he was traveling in Siberia, Kennan interviewed a co-conspirator in the plot. The details match those of Kropotkin's description of the jail break.
45. *Memoirs,* p. 371.
46. Dr. Crest Edward Veimer.
47. *Memoirs,* p. 371.
48. *Memoirs,* p. 372.
49. *Memoirs,* p. 373.
50. The Ministry of Justice eventually convened numerous committees that investigated the Kropotkin escape: Miller, 1976.
51. Other accounts claim Kropotkin had his beard shaved off at the apartment of Aleksandra Kornilova: Miller, 1976, p. 126.
52. *Memoirs,* p. 375. The advice was issued by Dr. Crest Edward Veimar; Kennan, 1912.

Notes to Chapter 4

1. Kropotkin, P. 1890 (1909). *Anarchist Morality.* "Freedom Office" 127 Ossulton St NW, p. 13.
2. *Memoirs,* p. 378.
3. The books were sent to the journal *Nature* by Sasha Kropotkin.
4. *Memoirs,* p. 382.
5. Kropotkin, P. 1880. An Appeal to the Young. La Revolte.
6. *Memoirs,* p. 398.
7. The Narodnaya Volya.
8. *Memoirs,* p. 449.
9. *Memoirs,* p. 451.
10. *Memoirs,* p. 451.

11. Woodcock, G., and I. Avakumovic. 1950. *The Anarchist Prince. A Biographical Study of Peter Kropotkin.* T.V. Boardman and Co., London. p. 194.
12. Shpayer-makov, H. 1987. The reception of Peter Kropotkin in Britian, 1886-1917. Albion 19:373-390.
13. Roylance-Kent, C. 1895. Anarchism: its origin and organization. *The Gentlemen's Quarterly,* April, p. 349.
14. *Memoirs,* p. 499.
15. Desmond, A. 1994. *Huxley: From Devil's Disciple to Evolution's High Priest.* Addison Wesley, New York, p. 6.
16. Huxley, T. 1909. *Autobiography and Selected Essays.* Houghton-Mifflin, Boston, p. 4.
17. Huxley, T.H. *Thoughts and Doings.* October 5, 1840 and November 22, 1840.
18. Huxley, T.H. 1901. *Autobiography and Selected Essays.* Houghton-Mifflin, Volume 1, p. 6.
19. Huxley 1901, p. 10.
20. Huxley, J. 1972. *T.H. Huxley's Dairy of the Voyage of the H.M.S. Rattlesnake.* Kraus Reprint. New York, p. vii-viii.
21. Hesketh, I. 2009. *Of Apes and Ancestors: Evolution, Christianity, and the Oxford Debate.* University of Toronto Press.
22. Huxley to Hooker, December 19, 1860.
23. The X-Club held its first meeting on November 3, 1864 at St. George's Hotel. This club was composed of the best minds in England, including Huxley, Hooker, Spencer and six others. In his creation of this society, Huxley had brought together a luminary group of thinkers, all with a devotion to Darwinism. The group made its mark directly with coups like the initiation of the journal *Nature.* The Xers also assumed an important role in promoting scientific education and free thinking; at a broader scale this manifested itself in the "invisible society" playing a significant part in making science part and parcel of Victorian society. More indiscreetly, its effects were felt via the placement of its members on important boards,

societies, and so forth. See: Barton, R. 1990. An influential set of chaps – The X-club and Royal Society Politics, 1864-85. British Journal for the History of Science 23:53-81; Barton, R. 1998. Huxley, Lubbock, and half a dozen others: professionals and gentlemen in the formation of the X Club, 1851-1864. Isis 89:410-444; Desmond, A. 2001. Redefining the X axis: "Professionals," "Amateurs" and the making of mid-Victorian biology: a progress report. Journal of the History of Biology 34:3-50; Waller, J. C. 2001. Gentlemanly men of science: Sir Francis Galton and the professionalization of the British life-sciences. Journal of the History of Biology 34:83-114; Barton, R. 2003. 'Men of science': Language, identity and professionalization in the mid-Victorian scientific community. History of Science 41:73-119.
24. Huxley, T. H. 1888. The struggle for existence: a programme. Nineteenth Century, 23, 161-180, p. 163-165.
25. Huxley, 1888, p. 169.
26. Huxley struck again five years later. On Thursday, May 18, 1893, a silver haired, ailing 68-year-old Huxley gave a lecture entitled "Evolution and Ethics" at Oxford University.
27. The Malthusian dilemma for Huxley was real, and England's only remedy was that her "produce shall be better than that of others." This economic, Adam Smith-like answer would then set the stage for the second of Huxley's requirements—"social stability." Such stability occurs when in society "the wants of its members obtain as much satisfaction as, life being what it is...may be reasonably." This would help break the grasp of the doom and gloom that faces us when we fail to combat our evolutionary predisposition to overreproduce.
28. Huxley, T.H. 1893. Evolution and Ethics, The Romanes Lecture, in Huxley, J. (1947). *Touchstone for Ethics, 1893-1943*. New York: Harper and Brothers. p. 92.
29. *Mutual Aid: A Factor of Evolution*, xlii.
30. Miller, 1976, p. 172.

31. *Mutual Aid: A Factor of Evolution*, p. 6.
32. *Mutual Aid: A Factor of Evolution*, p. 4.
33. *Mutual Aid: A Factor of Evolution*, p. 5.
34. Appendix VI, page 33 found in the unpublished appendices to Kropotkin's book, *Mutual Aid*. International Institute of Social History, Amsterdam.
35. And these are expanded even more in the appendix of Kropotkin's book, *Mutual Aid*.
36. Here, he cites the work of Syevertsoff; *Mutual Aid: A Factor of Evolution*, p. 8.
37. *Mutual Aid: A Factor of Evolution*, p. 10. At times, when referring to microorganisms, Kropotkin used "mutual support" rather than "mutual aid."
38. *Mutual Aid: A Factor of Evolution*, p. 10-11.
39. *Mutual Aid: A Factor of Evolution*, p. 11-12.
40. *Mutual Aid: A Factor of Evolution*, p. 14-15.
41. *Mutual Aid: A Factor of Evolution*, p. 20-21.
42. *Mutual Aid: A Factor of Evolution*, p. 20-23.
43. *Mutual Aid: A Factor of Evolution*, p. 25.
44. *Mutual Aid: A Factor of Evolution*, p. 25.
45. *Mutual Aid: A Factor of Evolution*, p. 51.
46. *Mutual Aid: A Factor of Evolution*, p. 51.
47. *Mutual Aid: A Factor of Evolution*, p. 51.
48. *Mutual Aid: A Factor of Evolution*, p. 75.
49. For more on the development of evolutionary ideas in Russia see: Rogers, J. A. 1973. The reception of Darwin's Origin of Species by Russian scientists. Isis 64:484-500; Rogers, J. A. 1974. Russian opposition to Darwinism in the Nineteenth Century. Isis 65:487-505; Todes, D. 1987. Darwin's Malthusian metaphor and Russian evolutionary thought. Isis 78:537-551; Todes, D. 1989. *Darwin without Malthus: The Struggle for Existence in Russian Evolutionary Thought*. Oxford University Press, New York; Glick, T. F., ed. 1974. *The Comparative Reception of Darwinism*. University of Texas Press, Austin.

50. Other leaders included Severtsov, Menzbir, Brandt, Filippov, Bekhterev and Bogdanov; Todes, 1989.
51. Peter Kropotkin in a letter to Marie Goldsmith, April 7, 1915.
52. Darwin 1859, p. 62-63.
53. Darwin, 1859, p. 63.
54. Huxley, T. H. 1873. *On the Origin of Species, or The Causes of the Phenomena of Organic Nature*. New York: D. Appleton, pages 123-124.
55. *Mutual Aid: A Factor of Evolution*, p. xxxvi.
56. *Mutual Aid: A Factor of Evolution*, p. 60.
57. Kropotkin, P. 1887. The scientific bases of anarchy. Nineteenth Century 21:238-252, p. 248.
58. Danilevskii, N. (ed.). 1869. *Rossiia i Evropa*. New York: Johnson Reprint Company, p. 146 (as cited in Todes, 1989).
59. Dokuchaev, V. V. 1953. Publichnye lektsii po pochvovedeniiu i sel'skomu khoziastvu (1890-1900). *Sochinenie*, 7, p.277 (as cited in Todes, 1989).

Notes to Chapter 5

1. Kropotkin, P. 1904. The ethical need of the present day. Nineteenth Century 56:207-226, p.225.
2. Kropotkin, P. 1890 (1909). Anarchist morality. "Freedom Office," 127 Ossulton St NW, p. 19.
3. Smith, A. 1759 (1853). *The Theory of Moral Sentiments*. Henry Bohn, London, p. 3.
4. Smith, A. 1759 (1853), p. 10. In the very same work, Smith wrote that "The administration of the great system of the universe ... the care of the universal happiness of all rational and sensible beings, is the business of God and not of man" p. 348. Kropotkin thought differently about the role of man and god in shaping human happiness.
5. Kropotkin, P. 1890 (1909). Anarchist morality. "Freedom Office," 127 Ossulton St NW, p. 16.
6. Kropotkin 1890 (1909), p. 17.

7. Kropotkin 1890 (1909), p. 17.
8. *Mutual Aid: A Factor of Evolution*, p. 76.
9. Hobbes, *Leviathan*.
10. Huxley, T. H. 1888. The struggle for existence: a programme. *Nineteenth Century, 23*, 161-180.
11. *Mutual Aid: A Factor of Evolution*, p. 77.
12. *Mutual Aid : A Factor of Evolution*, p. 77.
13. *Mutual Aid: A Factor of Evolution*, p. 82.
14. *Mutual Aid: A Factor of Evolution*, p. 90.
15. Mead, M. 1928. *Coming of Age in Samoa*. Morrow.
16. *Mutual Aid: A Factor of Evolution*, p. 95.
17. *Mutual Aid: A Factor of Evolution*, p. 101.
18. *Mutual Aid: A Factor of Evolution*, p. 104.
19. *Mutual Aid: A Factor of Evolution*, p. 117.
20. The Barbarian "invasions," Kropotkin believed, were a result of geological changes, most importantly desiccation, which led to large-scale societal migrations in search of water.
21. Since Hardin's classic paper, modern economists have obsessed over how to solve the free-rider problems that emerged from the use of commons. Hardin, G. 1968. The tragedy of the commons. Science 162:1243-1248.
22. *Mutual Aid: A Factor of Evolution*, p. 130.
23. *Mutual Aid: A Factor of Evolution*, p. 142.
24. *Mutual Aid: A Factor of Evolution*, p. 187.
25. *Mutual Aid: A Factor of Evolution*, p. 163.
26. *Mutual Aid: A Factor of Evolution*, p. 191.
27. *Mutual Aid: A Factor of Evolution*, p. 194-195.
28. *Mutual Aid: A Factor of Evolution*, p. 186.
29. *Mutual Aid: A Factor of Evolution*, p. 222.
30. Burkhardt, R. W. 1984. *The Zoological Philosophy of J.B. Lamarck in* R. W. Burkhardt, ed. Zoological Philosophy. University of Chicago Press, Chicago.
31. Lamarck, J.-B. 1809. *Zoological Philosophy*. Dentu, Paris.
32. Lamarck did not use the word evolve to describe how adaptations arose and spread within populations.

33. Lamarck, J.-B. 1801. *Systeme des Animaux sans Vertebres*. Agasse, Paris.
34. Darwin himself wrote, "Heaven forefend me from Lamarck's nonsense of a 'tendency to perfection,' 'adaptations from the slow willing of animals,' etc., but the conclusions I am led to are not widely different from his; though the means of change are wholly so." Letter from Darwin to Hooker, January 11, 1844. In *The Origin of Species*, however, Darwin did not completely rule out the inheritance of acquired characteristics.
35. Of course, today, much work has shown that Lamarck's theory of inheritance is fatally flawed, but Kropotkin can hardly be judged on modern findings.
36. Kropotkin, P. 1910. The theory of evolution and mutual aid. Nineteenth Century and After 67:86-107; Kropotkin, P. 1910. The direct action of environment on plants. Nineteenth Century and After 68:58-77; Kropotkin, P. 1910. The response of the animals to their environment. Nineteenth Century and After 68:856-867, 1047-1059; Kropotkin, P. 1912. Inheritance of acquired characteristics: theoretical difficulties. Nineteenth Century and After 71:511-531; Kropotkin, P. 1915. Inherited variation in animals. Nineteenth Century and After 78:1124-1144; Kropotkin, P. 1919. The direct action of environment and evolution. Nineteenth Century and After 85:70-89.
37. Kropotkin, P. 1919. The direct action of environment and evolution. Nineteenth Century and After 85:70-89, p. 89.
38. Giron, A. 2003. Kropotkin between Lamarck and Darwin. Asclepio LV:189-213.

Notes to Chapter 6

1. Kropotkin, What Geography Ought to Be, p. 942.
2. For example, in addition to *Mutual Aid: A Factor of Evolution*, Kropotkin wrote the following books and pamphlets (that sometimes were the length of short books): 1880. An Appeal

to the Young. La Revolte; 1885. What Geography Ought to Be. Nineteenth Century 18:940-956; 1892. *The Conquest of Bread*. Stock, Paris; 1880. Revolutionary Government. Freedom Press, London; 1887. *In Russian and French Prisons*. Ward and Downey, London; 1899. *Fields, Factories and Workshops*. Hutchinson, London; 1903. *The State: Its Historic Role*. Freedom Press, London; 1905. *Ideals and Realities in Russian Literature*. Alfred Knoph, New York; 1909. *The Great French Revolution*. William Heinemann, London; 1924. *Ethics; Origins and Development*. The Dial Press, New York. For the most comprehensive list of Kropotkin's books and pamphlets, see the appendix labeled "Kropotkiniana" in: Miller, M. A. 1976. Kropotkin. University of Chicago Press, Chicago.
3. Keltie, J. S. 1886. Report of the Proceedings of the Society in Reference to the Improvement of Geographical Education. John Murray. The report opens: "The object of the Council in appointing an Inspector of Geographical Education is to obtain fuller information than they now possess regarding the position and methods of Geographical education in this country and abroad. I. As regards the United Kingdom. A. The Inspector will ascertain by means of correspondence or by actual inspection (1) the extent to which Geography of any kind is taught in our Universities and public schools and what special rewards are offered for proficiency in it (2) the actual subjects taught under that name and the comparative time allotted to them (3) the methods and appliances used in teaching these subjects (4) the attitude of teachers and professors with regard to Geography generally (5) the value allotted to Geography in University examinations and with what subjects it is united."
4. Keltie, J. S. 1921. Obituary. Prince Kropotkin. The Geographical Journal 57:316-319.
5. Kearns, G. 2004. The political pivot of geography. Geographical Journal 170:337-346.

6. Kropotkin envisioned the "exchange, between schools, of correspondence on geographical subjects, and of their natural science collections," using a model that had been developed by the Agassiz Association. Kropotkin, What Geography Ought to Be, p. 952. Also see: Ballard, H. 1887. History of the Agassiz Association. Science 9:93-96.
7. Kropotkin, 1885. What Geography Ought to Be. Nineteenth Century 18:940-956, p. 940.
8. Kropotkin, What Geography Ought to Be, p. 941.
9. Kropotkin, What Geography Ought to Be, p. 942.
10. From "Prisons and their Moral Influences on Prisoners," December 20, 1877. Pages 220-235 in Baldwin, R., ed. 1971. Kropotkin's Revolutionary Pamphlets. Dover Press, NY.
11. Kropotkin, Prisons and their Moral Influences on Prisoners, p. 221.
12. Kropotkin, Prisons and their Moral Influences on Prisoners, p. 235.
13. Kropotkin, Prisons and their Moral Influences on Prisoners, p. 235.
14. He had already read some on the subject in his sister's library when he was allowed leave from the Corps of Pages. Peter's first significant publication on this subject was Kropotkin, P. 1889. The Great French Revolution and its lessons. Nineteenth Century 1889: 838-851. That same year, he published a similar piece, "Le Centeraire de La Revolution," in La Revolte.
15. Though he did make one speech in Paris in 1877.
16. Originally published in French, and later translated to English. I will be using the English version for page citations: Kropotkin, P. 1909. *The Great French Revolution*. William Heinemann, London.
17. *The Great French Revolution,* p. 97.
18. *The Great French Revolution,* p. 574.
19. *The Great French Revolution,* p. 580. For more on this, see Chapter 12 of: Morris, B. 2003. *Kropotkin: The Politics of Community*.

Humanity Books Amherst, NY, and Woodcock, G. 1990. Kropotkin's The Great French Revolution. Pp. 1-18 *in* D. Roussopoulos, ed. The Anarchist Papers, Volume 3. Black Rose.

20. Kropotkin, 1887. The scientific bases of anarchy, Nineteenth Century 21:238-252, p. 238.
21. Kropotkin, The scientific bases of anarchy. p. 238.
22. Kropotkin, 1908, *Modern Science and Anarchism*. Mother Earth Publisher, p. 57.
23. Kropotkin, Modern Science and Anarchism, p. 43. Darwin's actual description of the phenomena in *The Descent of Man* is as follows: "It must not be forgotten that although a high standard of morality gives but a slight or no advantage to each individual man and his children over the other men of the same tribe, yet that an increase in the number of well-endowed men and an advancement in the standard of morality will certainly give an immense advantage to one tribe over another. A tribe including many members who, from possessing in a high degree the spirit of patriotism, fidelity, obedience, courage, and sympathy, were always ready to aid one another, and to sacrifice themselves for the common good, would be victorious over most other tribes; and this would be natural selection. At all times throughout the world tribes have supplanted other tribes; and as morality is one important element in their success, the standard of morality and the number of well-endowed men will thus everywhere tend to rise and increase." Scholars still debate the extent to which Darwin thought this a general principle. Ruse, M. 1980. Charles Darwin and group selection. Annals of Science 37:615-630.
24. Kropotkin, *The Conquest of Bread*, p. 49.
25. Kropotkin, P. 1887. The scientific bases of anarchy, p. 238.
26. *The Conquest of Bread*, p. 44.
27. *The Conquest of Bread*, p. 42.
28. *The Conquest of Bread*, p. 46.

29. *The Conquest of Bread*, p. 57
30. *The Conquest of Bread*, p. 86.
31. *Memoirs,* p. 399.
32. *The Conquest of Bread,* p. 125.
33. Kropotkin, 1908, Modern Science and Anarchism, p. 13-14.
34. Kropotkin, P. 1924. *Ethics: Origins and Development*. The Dial Press, New York, p. 12.
35. Kropotkin, 1924. *Ethics*, p. 15. Also see, Kropotkin, 1904. The Ethical Need of the Present Day. Nineteenth Century 56:207-226; Kropotkin, 1905. The morality of nature. Nineteenth Century and After 58:865-883.
36. Kropotkin, 1890, Anarchist Morality, p. 18
37. Kropotkin, 1890, Anarchist Morality, p. 8.
38. Kropotkin, 1890, Anarchist Morality, p. 11.

Notes to Chapter 7

1. Flynne, E. G. 1973. *The Rebel Girl: An Autobiography*. Publications International, New York. Also see Camp, H. 1995. *Iron in Her Soul: Elizabeth Gurley Flynn and the American Left*. Washington State University.
2. Who was known as Stepniak.
3. Avrich, P. 1980. Kropotkin in America. International Review of Social History 25:1-34. Peter had long been an admirer of the United States' Declaration of Independence and Constitution. He was also fond of the Union movement. American capitalism, on the other hand, he despised. When socialist union leader Eugene Debbs was jailed for his role in the Pullman car strike, Peter even sent him a signed copy of his books to show his comradery.
4. Flynne, E. G. 1973. *The Rebel Girl: An Autobiography*. Publications International, New York. Also see Camp, H. 1995. *Iron in Her Soul: Elizabeth Gurley Flynn and the American Left*. Washington State University.

5. Goldman, E. 1931. *Living My Life*. De Capo Press, New York. P. 509.
6. Report of the Meeting of the British Association for the Advancement of Science, Volume 67, J. Murray, 1898, pp 648 and 722.
7. Dr. George Mercer Dawson.
8. Kropotkin, P. 1898. Some of the resources of Canada. Nineteenth Century 43:494-513, p. 494.
9. Kropotkin, 1898, p. 495.
10. Kropotkin, 1898, p. 498.
11. Kropotkin, 1898, p. 494.
12. Kropotkin, 1898, p. 503.
13. Kropotkin, 1898, p. 503.
14. Kropotkin to Georg Brandes, June 28, 1898.
15. Kropotkin, 1898, p. 495.
16. At the Lattimer coal mine.
17. Kropotkin, La Tuerie de Hazelton, in *Les Temps Nouveaux*, October 9-15, 1897. In English, La Tuerie de Hazelton means "the killing at Hazelton."
18. J. Most, October 30, 1897, "Peter Krapotkin" in *Frieheit* (Buffalo).
19. For more Labadie, see the Labadie Collection housed at The University of Michigan, http://www.lib.umich.edu/labadie-collection.
20. New York Herald, October, 24, 1897.
21. Purchase, G. 1996. *Evolution and Revolution: An Introduction to the Life of Peter Kropotkin*. Jura Books.
22. *New York Times*, October 24, 1897.
23. *New York Times*, October 25, 1897.
24. Knight-Smith, H. 1898. *The History of the Lowell Institute*. Lamson, Wolffe and Company.
25. Robert Erskine Ely and Walter Hines Page.
26. *New York Times*, March 28, 1901 "Russians meet and denounce Czar;" *New York Times*, March 30, 1901 "Prince Kropotkin on

Russia's condition;" *New York Times*, March 30, 1901 "Russian Nihilists Appeal."
27. *New York Times*, March 31, 1901 "Turgueneff and Tolstoi."
28. *New York Times*, April 1, 1901 "Prince Krapotkin talked on anarchy."
29. Avrich, 1980, p. 24.
30. Berkman, A. 1912. *Prison Memoirs of an Anarchist*. Mother Earth Publishing, p. 442.
31. Addams, J. 1910. *Twenty Years at Hull-House: with Autobiographical Notes*. MacMillan.
32. Addams, 1910, p. 402-403.
33. Havel, H., "Kropotkin the Anarchist," Mother Earth, December *1912*.
34. Kropotkin to Guillaume, December 12, 1901.

Notes to Chapter 8

1. Russell, B. 1919. *Proposed Roads to Freedom*. Henry Holt, New York, p. 192.
2. Woodcock, G., and I. Avakumovic. 1950. *The Anarchist Prince. A Biographical Study of Peter Kropotkin*. T.V. Boardman and Co., London, p. 292.
3. Kropotkin, P. 1899. *Fields, Factories and Workshops*. Hutchinson, London, p. 78.
4. Kropotkin, P. 1908. *Modern Science and Anarchism*. Mother Earth, p. 53.
5. Kropotkin to Marie Goldsmith, Jan 7, 1915.
6. Kropotkin, P. 1905. *The Revolution in Russia*. Nineteenth Century, December 1905:865-883.
7. Kropotkin had also been in France in 1905 to visit an old friend, James Guillaume.
8. Woodcock, 1950, pp. 366-367.
9. Such as allowing the people more control over electing representatives to the Duma (State Council).

10. Ascher, A. 2004. *The Revolution of 1905: a Short History*. Stanford: Stanford University Press.
11. Kropotkin, P. 1909. *The Terror in Russia*, London.
12. Kropotkin, P. 1909. *The Terror in Russia*, London, p. 32.
13. Kropotkin, P. 1909. *The Terror in Russia*, London, p. 75. As he wrote *The Terror in Russia*, Kropotkin the evolutionary theoretician was also at work, creating new models for mutual aid—models that might make mutual aid in humans easier to conceptualize. It was during this period (from 1909 forward) that Peter published his series of articles on Lamarckian evolution and mutual aid. Lamarck's views on the inheritance of acquired characteristics allowed for mutual aid to emerge not just over evolutionary time, but over the course of human timeframes, as Peter hoped it would in Russia and everywhere else. See Chapter 6 of this book. Kropotkin, P. 1910. The direct action of environment on plants. Nineteenth Century and After 68:58-77; 1910. The response of the animals to their environment. Nineteenth Century and After 68:856-867, 1047-1059; 1912. Inheritance of acquired characteristics: theoretical difficulties. Nineteenth Century and After 71:511-531; 1915. Inherited variation in animals. Nineteenth Century and After 78:1124-1144; 1919. The direct action of environment and evolution. Nineteenth Century and After 85:70-89.
14. Kropotkin, P. 1910. Anarchism. Encyclopedia Britannica.
15. Kropotkin, P. 1910. Anarchism. Encyclopedia Britannica.
16. Led by Lvov, and then Kerensky.
17. *New York Times*, March 17, 1917.
18. Shub claims a crowd of 60,000, but this number has never been confirmed. Shub, D. 1953. Kropotkin and Lenin. The Russian Review XII:227-234.
19. Shub, 1953.
20. Shub, 1953, p. 233.
21. Woodcock and Avakumovic, 1950, p. 380.
22. Miller, 1976 p. 240.

23. Woodcock and Avakumovic, 1950, p. 432.
24. Emma Goldman in Ishill, J. 1923. *Peter Kropotkin, The Rebel, Thinker and Humanitarian.* The Free Spirit Press, Berkeley Heights, NJ.
25. *New York Times,* October 15, 1920.
26. *New York Times,* January 29, 1921 and February 3, 1921.
27. *New York Times,* February 12, 1921.
28. *Memoirs,* p. 25.

Notes to Epilogue

1. In Ishill, J. 1923. *Peter Kropotkin, The Rebel, Thinker and Humanitarian.* The Free Spirit Press, Berkeley Heights, NJ.

INDEX

Addams, Jane, 80
Alexander II (Czar), x, 7-9, 11, 35
Altruism (see mutual aid)
American Civil Liberties Union, 73
Amur region of Siberia, 8-20
Atlantic Monthly (magazine), 37, 78, 79

Berkeley Lyceum, 80
Berkman, Alexander, 80
British Association for the Advancement of Science, 39, 73, 74

Catherine the Great (Empress), 9, 24
Chickering Hall, 77, 78
Clairvaux Prison, 36, 64
Cooperation (see mutual aid)
Cooper Union College, 78
Corps of Pages, 4-9
Czolgosz, Leon, 82

Darwin, Charles, x-xii, 6, 13, 14, 38, 39, 42, 46-49, 60, 61, 67, 70
Davis, Varina, 80
Dostoevsky, Fyodor, 11

Ely, Robert, 80

Flynn, Elizabeth Gurley, 72, 73

Goldman, Emma, 73, 82, 90, 91
Gray, Asa, 77
Great Fire of St. Petersburg, 9

Hale, Edward Everett, 79
H.M.S. Beagle, 9
H.M.S. Rattlesnake, 38
Hobbes, Thomas, 40, 53, 55, 60
Holmes, Oliver Wendell, 78
Homestead Steel strike, 80
Hull House, 80, 81
Humboldt, Alexander von, x, 4, 8, 16
Huxley, Thomas Henry, 37-41, 47, 49, 53, 55, 60, 70

James, William, 78
Jura Federation, 19, 34, 35

Keltie, J. Scott, 34, 63
Kerensky, Alexander, 89
Kessler, Karl Fedorovich, 46

Index

Knowles, James, 41
Kropotkin, Alexander (Sasha, brother), 2-6, 12, 13, 15, 19, 27, 28
Kropotkin, Alexander (father), 2, 3, 5, 8, 9, 19
Kropotkin, Ekaterina Sulima (mother), 2
Kropotkin, Elisabeth Karadino (step-mother), 2
Kropotkin, Nicholas (brother), 2
Kropotkin, Peter, all throughout
Kropotkin, Sophie Ananiev (wife), 35
Kropotkin, Yelina (sister), 2, 5, 28

Lamarck, Jean-Baptiste, 59-61, 66
Lenin, Vladimir, 89-91
Le Révolté (newspaper), 35
Leviathan (see, Hobbes, Thomas)
Lowell Institute, 77-79
Lyell, Charles, 77

McKinley, William, 82
Malthus, Thomas, 47-49, 53, 55, 61
Manifesto of 16, 88
Marx, Karl, 84, 85
Mavor, James, 74-76, 78, 81
Mutual aid, x-xiii, 12-21, 37, 41-72, 75, 77, 81, 84-94

Nineteenth Century (magazine), 37, 39, 41, 42, 60, 63, 66, 75
Norton, Charles Eliot, 78

October 1917 revolution, 89-91
Odd Fellows Temple, 77
Old Equerries section of Moscow, xiii, 1
Origin of Species, The, 6, 47, 59

Paris Commune, 19
Peter and Paul Prison, 25-32, 36, 37, 64, 73, 89
Potter, Beatrix, 39
Proudhon, Pierre-Joseph, 11, 12, 17

Rurik Dynasty, xiii, 2
Russell, Bertrand, 83
Russian Geographical Society, 20-24

Smith, Adam, 49-53
Spencer, Herbert, 19

Tammany Hall, 80
Tchaykóvsky Circle, 22-24
Theory of Moral Sentiments, The, 49-53
Trotsky, Leon, 90

Voltaire (Francois Marie Arouet), 5

Wallace, Alfred Russell, 78
Washington, Booker T., 80
Wealth of Nations, The (see Smith, Adam), 51
Wilberforce, Samuel, 39
Wilde, Oscar, ix
Women's Industrial Club of Cambridge, 78

X-Club, 39

Zoological Philosophy (see Lamarck, Jean-Baptiste)

Made in the USA
Lexington, KY
02 February 2012